T0335924

Synchronization and Control of Chaos

Synchronization and Control of Chaos

An Introduction for Scientists and Engineers

J. M. González-Miranda
University of Barcelona, Spain

ICP Imperial College Press

Published by

Imperial College Press
57 Shelton Street
Covent Garden
London WC2H 9HE

Distributed by

World Scientific Publishing Co. Pte. Ltd.
5 Toh Tuck Link, Singapore 596224
USA office: 27 Warren Street, Suite 401-402, Hackensack, NJ 07601
UK office: 57 Shelton Street, Covent Garden, London WC2H 9HE

British Library Cataloguing-in-Publication Data
A catalogue record for this book is available from the British Library.

SYNCHRONIZATION AND CONTROL OF CHAOS

ISBN-13 978-1-86094-488-8
ISBN-10 1-86094-488-4

Printed in Singapore

A mi padre, Alfonso, y
a mi madre, María Angeles.

Preface

Over the last fifteen years an important research activity has been devoted to the dynamics of coupled and driven chaotic oscillators. This is still an active field of research; however, a considerable body of knowledge has already been established. This is mainly, but not exclusively, in the form of research papers in physics and mathematics journals. These results are of interest for scientific analysis and explanation in all scientific disciplines, as well as for possible application in medicine and engineering. The purpose of this book is to provide a systematic and broad account of that research for a wide audience. This will be achieved by a selection of the most basic theoretical results, as well as experiments and applications to be presented at a mathematical level accessible for people working in non-hard sciences. This, however, does not exclude physicists and mathematicians looking for an introduction to the field.

Resonance and normal modes of vibration are well known classical phenomena observed in driven and coupled linear harmonic oscillators, which have a great relevance in all the natural sciences and in engineering. Since the last quarter of the twentieth century the study of chaotic oscillators has emerged as an object of great interest in physics and mathematics. In view of the importance that the classical results for the dynamics of driven, or coupled harmonic oscillators, has achieved in science and technology, the question of which phenomena emerge when chaotic oscillators are coupled or somehow driven or perturbed has been, and is of great interest. The most relevant phenomena studied until now are the synchronization and the suppression of chaos. This book is aimed to provide a brief account of these investigations. The approach used considers different schemes of driving or coupling. For each of these set-ups, the phenomena that occur are studied, and experiments that show that these phenomena are observable in the real world are discussed.

The first two chapters, which deal with the theory of oscillators, harmonic and chaotic, are aimed to provide the reader with basic concepts on dynamical systems and chaos theory needed to follow the rest of the book. In the next five chapters, the different scenarios which emerge when chaotic oscillators are externally driven or mutually coupled are studied. These include a variety of forms of chaos synchronization, which are systematically defined and characterized; as well as several conditions, and procedures which lead to taming chaos; i.e., to turn the chaotic motion to periodic. Moreover, throughout these seven chapters, by means of examples borrowed from different disciplines, the multidisciplinary nature of the subject will be illustrated. In the last chapter, a general vision of the field of synchronization and control of chaos is given by means of a summary, and examples of application of this body of knowledge in science and technology are presented to sustain a discussion of perspectives for the future.

The didactical approach followed throughout the book combines tutorial and review techniques. For each scientific issue studied, the basic concepts and techniques are explained in detail and immediately exemplified by means of numerical simulations made with simple, but significant models. This is followed by the presentation of a brief review and discussion of experimental realizations and numerical simulations. The former will show that the theory introduced is meaningful in the real world, and the latter will expand the scope of the theory and examples previously posed. All the above is provided together with a bibliography aimed to allow the reader to probe deeper into the particular issues that especially interest him or her.

I have been working on coupled and driven chaotic oscillators for the last ten years. The view presented in this book, has been shaped during my experience with the development of my own research and the interaction with many colleagues, whom I want to acknowledge and thank. I also gratefully acknowledge the financial support from DGI which allowed the scientific activity that has made this book possible.

Jesús Manuel González Miranda

Barcelona, Spain.
December 2003

Contents

Chapter 1

Dynamics of Coupled and Driven Harmonic Oscillators

This chapter is an introduction aimed to insert the body of the book in the context of the physical sciences, and to introduce notation and basic concepts to be used in the remainder of the book. These include the notions of equilibria, dissipation and forcing, as well as that of dynamical systems and the phase space. This will be done by means of a reminder of the study of the harmonic oscillator and of two basic, fundamental and fruitful model-systems of elementary mechanics: the periodically forced harmonic oscillator and the system of two coupled harmonic oscillators. At the end of the chapter nonlinearity and chaos will be introduced through the example of simple models that generalize the harmonic oscillator.

1.1 The harmonic oscillator

Oscillatory phenomena pervade the natural sciences, and the engineering world. Pulsating stars are studied in astrophysics, while in astronomy the motions of the planets in their orbits have an oscillatory nature. This, in the case of Earth, results in the periodic repetition of the four seasons year after year. Earthquakes which are vibrations of the Earth's surface, and long term variations of the Earth's magnetic field are examples of oscillatory phenomena in the Earth sciences. Oscillatory behaviors are often found in the life sciences too; these include the circadian rhythms, the beats of the heart, and the oscillations of the membrane potential in the axons of the neurons, among many others. It is common to find in chemistry oscillating chemical reactions in which the amounts of the reactants change periodically with time, as well as vibrations in the motions of the atoms that constitute the molecules. Physics is full of oscillatory phenomena too; electromagnetic fields whose intensity changes periodically with time, known as electromagnetic waves, atomic vibrations in solid state physics, and modes of oscillation of the atom nucleus are examples of this kind of phenomena. Electrical and mechanical oscillators, as well as vibrations in structures are everyday elements in the world of engineering.

The study of the dynamical behavior of oscillating systems, oscillators for short, is therefore a central issue in physics and in mathematics. Then, these sciences provide basic and general results that found major applications not only in physics, but also in all the other branches of science, as well as in technology. The harmonic oscillator is the simplest, and more fundamental theoretical model of oscillatory phenomena; therefore, this book, which deals with some recent developments in the physics of the kind of oscillators known as chaotic oscillators, will start with a review of the main results for the harmonic oscillator.

1.1.1 *The free harmonic oscillator*

When, in science and technology, a system is a free harmonic oscillator, the result of the observation of its state, as given by an appropriate measured variable, x, changes with time, t, according to a law like

$$x(t) = A \sin(\omega_0 t + \delta), \tag{1.1}$$

which, as shown in Fig. 1.1(a), is characterized by a series of peaks and valleys, of height A and depth $-A$, which repeat themselves in time with a period $T = 2\pi/\omega_0$. This simple harmonic oscillation describes, at least approximately, a wealth of cases found in practice.

The dynamical behavior of the harmonic oscillator can easily be studied by means of simple mechanical systems like a pendulum in the gravitational field under small oscillations [Fig. 1.1(b)], or a mass attached to a spring and moving on a frictionless surface [Fig. 1.1(c)]. The observable x, in the case of the pendulum is the angle which measures the deviation from the vertical line, $x = \theta$; while for the mass attached to a spring, x is the deviation from the position where the spring is relaxed (not stretched, nor compressed), which is assumed to be $x = 0$. According to Newton's laws of motion, in these two cases $x(t)$ has to obey the following dynamical law

$$\frac{d^2x}{dt^2} + \omega_0^2 x = 0, \tag{1.2}$$

which is a linear second order differential equation, whose general solution is precisely Eq. 1.1. For the pendulum it is $\omega_0^2 = g/L$, with g the acceleration of gravity, and L the length of the pendulum. For the spring model it is $\omega_0^2 = k/m$, with m the value of the mass, and k the strength of the restoring force of the spring, $F = -k \cdot x$.

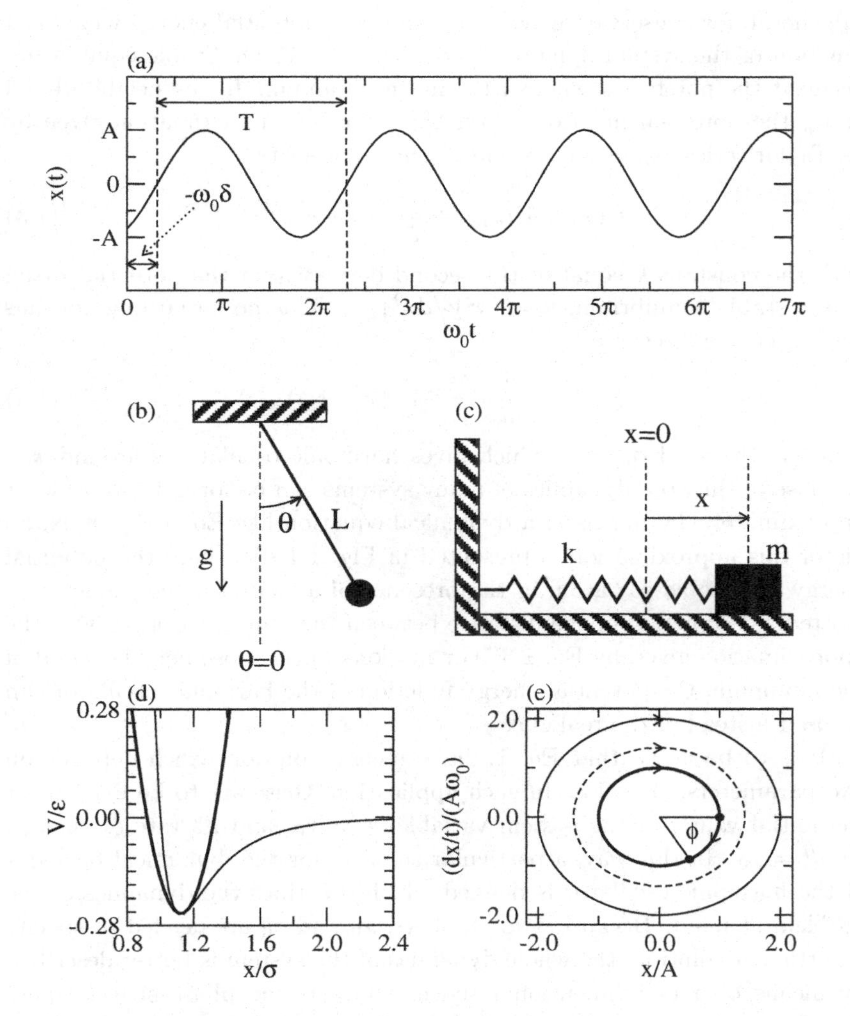

Fig. 1.1 (a) Time series for a free harmonic oscillator. Two examples: (b) the pendulum, and (c) the mass-spring system. (d) Approximation of an equilibrium for the Lennard-Jones 12-6 potential, $V(x) = \varepsilon\left[(\sigma/x)^{12} - (\sigma/x)^6\right]$ (thin line), by means of a Taylor series, Eq. 1.3, (thick line). (e) Phase space trajectories for initial conditions: $x_0 = A$ (thick line), $x_0 = \sqrt{2}A$ (dashed line), $x_0 = 2A$ (thin line), and $\dot{x}_0 = 0$ in all cases.

The pendulum and the spring are two models that allow easy visualization and understanding of harmonic oscillations. The results given by Eq. 1.1 and Eq. 1.2, however, are far-reaching as they model the case when the dynamics of a system occurs in close proximity to a stable equilibrium. In general, for conservative systems, there is a potential energy which is a function of the system dynamical variable, $V = V(x)$. Stable equilibrium occur at the points x_m where $V(x)$ has a minimum. In the neighborhood of x_m the potential function allows the following approximation given by its Taylor series expansion around x_m up to order two

$$V(x) \approx V(x_m) + \frac{1}{2}k \cdot (x - x_m)^2, \tag{1.3}$$

with the constant k equal to the second derivative of the potential taken at the stable equilibrium, $k = \left(d^2V/dx^2\right)_{x_m}$. This potential function has associated the force

$$F = -\frac{dV}{dx} = -k \cdot (x - x_m), \tag{1.4}$$

which is the kind of force which gives harmonic oscillations around x_m. Because of this, the dynamics of many systems can be formulated, at least approximately, by means of a dynamical equation like Eq. 1.2. An example of this approximation is presented in Fig. 1.1(d), where the potential energy of a noble gas atom in the presence of a second noble gas atom is plotted as a function of the distance between the atoms, together with the approximation given by Eq. 1.3. For motions in the close neighborhood of the minimum, the potential energy function of the harmonic oscillator can be used instead of the real one.

It is to be noted that Eq. 1.1 is a general solution which depends on two parameters, A and δ. In each application, these are to be fixed to fit the initial values of the system variable, $x = x_0$, and its rate of change, $dx/dt = \dot{x}_0$. In this way, a particular solution for the dynamical behavior of the harmonic oscillator is defined which describes the dynamics of the problem at hand. Because, two numbers, x_0 and $\dot{x}_0$ are needed to specify a particular solution, the whole dynamics of the system is better described by means of a two-dimensional space, know as the phase space, whose coordinates are x and $\dot{x} = dx/dt$. Each particular solution in the phase space appears [Fig. 1.1(e)] as a trajectory which has the shape of an ellipse centered at the point that corresponds to the stable equilibrium, whose coordinates are $(x_m, 0)$. The actual state of the system is given by a point on this ellipse, know as phase space point or representative point, whose dynamics is then given by a rotation along the ellipse with the angular velocity ω_0. This is called the natural frequency of the harmonic oscillator, and defines the actual position of the phase space point by means of the angle $\phi = \phi(t)$, which obeys the dynamical equation $d\phi/dt = \omega_0$, and it is known as the phase of the harmonic oscillator.

The use of dimensionless units for the position, velocity and time in these plots is to be noted. This is a common and convenient practice in the literature of dynamical systems.

1.1.2 *Damped harmonic oscillator*

An element which is relevant in most practical cases is dissipation. The simple examples presented above, the pendulum and the mass-spring system, are idealizations that can be approached in the laboratory when friction is small, and the systems are observed for intervals of time short enough as to have a negligible amount of energy lost by dissipation. However, the common case is when the energy of the system is lost at a certain rate because of the presence of friction, or some other dissipative effect. Dissipation is meant as an effect which opposes to the change of x, and in a simple and effective approximation, it is assumed that it behaves like a friction force proportional to $\dot{x}$. In this approximation, the differential equation for the dynamical variable, $x(t)$, of the harmonic oscillator becomes

$$\frac{d^2x}{dt^2} + 2\gamma\frac{dx}{dt} + \omega_0^2 x = 0, \tag{1.5}$$

where the new positive parameter, γ, measures the intensity of the dissipation.

The general solution of this dynamical equation can be obtained by means of standard techniques of solution of linear ordinary differential equations [Goldstein et al. (2002); Kibble and Berkshire (1996)]. The result is such that $x(t)$, for all $\gamma > 0$, decays at an exponential rate to the position of the equilibrium point, which is assumed to be zero in the formulation of Eq. 1.5. The details of the decaying law are dependent on the relation between γ and ω_0, as stated by the following expression:

$$x(t) = \begin{cases} Ae^{-\gamma t}\cos(\omega_\gamma t + \delta), \text{ for } \omega_0^2 > \gamma^2, \\ (C_1 + C_2 t)e^{-\gamma t}, \quad \text{for } \omega_0^2 = \gamma^2, \\ C_1 e^{-\gamma_+ t} + C_2 e^{-\gamma_- t}, \text{ for } \omega_0^2 < \gamma^2, \end{cases} \tag{1.6}$$

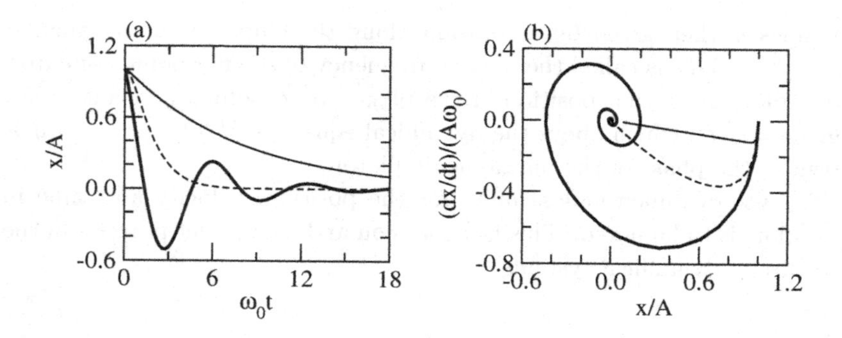

Fig. 1.2 (a) Time series, and (b) phase space plots, for a dumped harmonic oscillator in an under-dumped case, $\gamma = \omega_0/4$, (thick line), an over-dumped case, $\gamma = 4\omega_0$, (thin line), and the critical case $\gamma = \omega_0$, (dashed line).

with $\omega_\gamma^2 = \omega_0^2 - \gamma^2$, and $\gamma_\pm = \gamma \pm \sqrt{-\omega_\gamma^2}$. The constants, A, δ, C_1 and C_2 define the particular solutions for given initial conditions as before. The first case corresponds to low dumping and gives an oscillating decay to the stable equilibrium, the third case corresponds to strong dumping and gives a fast monotone decay, while the second case is a critical one that marks the transition between the other two. These three types of time evolutions are illustrated in Fig. 1.2(a), for $x(t)$, and in Fig. 1.2(b) for the whole dynamical behavior in the phase space.

An important property of the harmonic oscillator which has to be stressed is linearity. This means that the dynamical equations are such that if $x_a(t)$ and $x_b(t)$ are two independent solutions of Eq. 1.2 or Eq. 1.5, then any linear combination of them $x(t) = c_a \cdot x_a(t) + c_b \cdot x_b(t)$, with c_a and c_b real numbers is also a solution. This property is used in the theory of linear ordinary differential equations to construct techniques of solution.

1.2 Driven and coupled harmonic oscillators

Many systems allowing a description as harmonic oscillators are well modelled by Eq. 1.2 and Eq. 1.5; however, in practical cases, these systems are not usually isolated. Instead, they interact with the environment, or with other oscillators. Therefore, the study of the dynamics of driven and coupled oscillators are issues of major importance.

1.2.1 *Periodically driven harmonic oscillator: resonance*

A relatively simple, though very interesting case, is when the external action is periodic. That is, the case of a damped harmonic oscillator acted on by an external periodic force, whose dynamical behavior, $x(t)$, is described by the linear second order differential equation

$$\frac{d^2x}{dt^2} + 2\gamma\frac{dx}{dt} + \omega_0^2 x = A_1 \sin(\omega_1 t), \tag{1.7}$$

where the term in the second member is the external periodic force, which has a frequency ω_1, and whose strength is measured by the amplitude A_1.

Again, the theoretical study of this case can be obtained as an application of the theory of linear ordinary differential equations. The general solution that is obtained is

$$x(t) = x_D(t) + \frac{A_1 \sin(\omega_1 t + \beta)}{\sqrt{\left(\omega_0^2 - \omega_1^2\right)^2 + (2\gamma\omega_1)^2}}, \tag{1.8}$$

with $\beta = -\arctan\left[(2\gamma\omega_1)/\left(\omega_0^2 - \omega_1^2\right)\right]$, and $x_D(t)$ given by Eq. 1.6.

The main features of the dynamical behavior that is described by Eq. 1.6 are illustrated in Fig. 1.3(a). For t small, there is a transitory behavior whose details are determined by the relation between γ and ω_0, and the particular values of the initial conditions, x_0 and $\dot{x}_0$. This decays exponentially with time because this is what $x_D(t)$ does. After this transitory has died of, $x(t)$ is given by

$$x(t) = A \sin(\omega_1 t + \beta), \tag{1.9}$$

which is an harmonic-like oscillation, which has the frequency of the driving force. The amplitude

$$A_S(\omega_1) = \frac{A_1}{\sqrt{\left(\omega_0^2 - \omega_1^2\right)^2 + (2\gamma\omega_1)^2}}, \tag{1.10}$$

and the phase angle, β, are determined not by the initial conditions, but by the parameters, ω_0, γ, A_1 and ω_1, that describe the oscillator. This periodic oscillation is known as the stationary solution because it is the one that lasts for large t. One of its relevant features is that it has lost memory of the initial conditions because of dissipation. The dynamics of stationary solutions in phase space as presented in Fig. 1.3(b) shows that, after a transitory evolution, the system settles down in a motion on a cycle which has the shape of an ellipse, similar to those presented in Fig. 1.1(e),

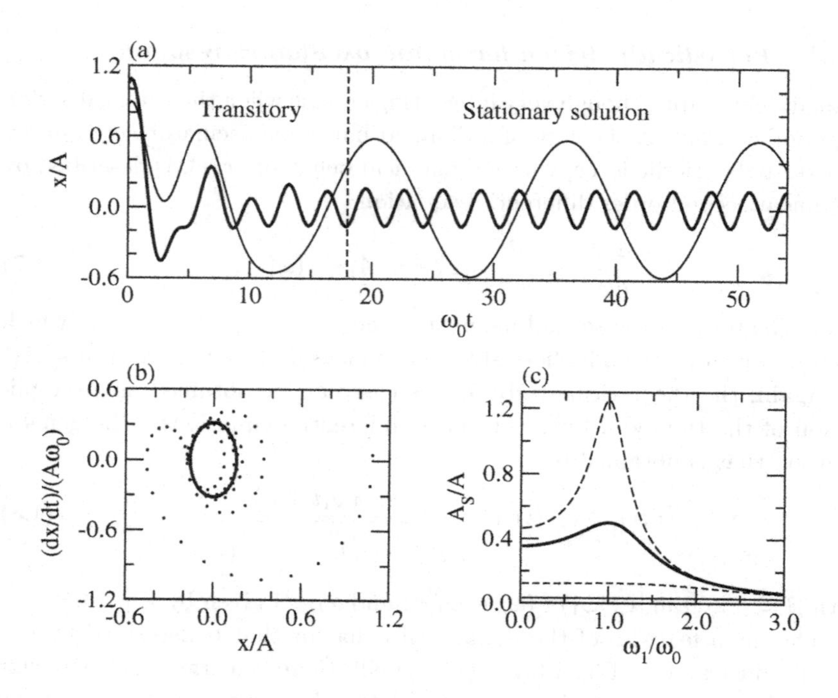

Fig. 1.3 (a) Time series for an under-dumped harmonic oscillator driven by a periodic force of amplitude $A_1 = A/2$, and frequencies $\omega_1 = 2\ \omega_0$ (thick line), and $\omega_1 = 2\ \omega_0/5$ (thin line), (b) Phase space plot for the under-dumped harmonic oscillator driven by a periodic force ($A_1 = A/2$, and $\omega_1 = 2\ \omega_0$). (c) Dependence of the amplitude of the stationary solution on the frequency of the applied force, for different values of the dumping: $\gamma = 4\omega_0$ (thin line), $\gamma = \omega_0$ (short dashed line), $\gamma = \omega_0/4$ (thick line), and $\gamma = \omega_0/10$ (long dashed line).

which is called a limit cycle. The change in $\omega_1$has a remarkable effect in the amplitude of the oscillations: a systematic study of the dependence of A_S on ω_1 shows [Fig. 1.3(c)] a second relevant property of the stationary solution: when $\omega_1 = \omega_0$ the amplitude reaches a maximum, which becomes more pronounced when the dissipation decreases. This phenomenon, which is understood just by an inspection of Eq. 1.10, is known as resonance, and plays an important role in the understanding of many phenomena and techniques in science and technology.

Resonance curves like those plotted in Fig. 1.3(c), which are given by equations like Eq. 1.10, describe a phenomenon which is not limited to mechanical oscillators. In fact they describe, at least qualitatively, the response of any linear oscillator that is driven by an external force which varies sinusoidally with time. This is why resonance is a widespread phenomenon. For example, in electrical engineering, electrical oscillators can be constructed by a simple electric circuit in which a capacitor, of capacity C, a linear resistance, of resistance R, and an inductor, of inductance L, are set in series and powered by an *ac* voltage source which provides a sinusoidal voltage of amplitude V_0 and frequency ω_1. The theory of electric circuits gives the following differential equation for the time evolution of the time variation of the electric charge, q, in the capacitor of such circuit

$$\frac{d^2q}{dt^2} + \frac{R}{L}\frac{dq}{dt} + \frac{1}{LC}q = \frac{V_0}{L}\sin(\omega_1 t), \tag{1.11}$$

which is formally identical to Eq. 1.7, and therefore must have identical solutions. This means that the RLC circuit just described must display, and in fact displays, phenomena of electrical resonance when ω_1 is varied in the neighborhood of $\omega_0 = 1/\sqrt{LC}$, being the resonance peak more pronounced for small values of $\gamma = R/2L$.

Another example is nuclear magnetic resonance (NMR), an experimental technique used in the natural sciences, medicine and engineering. In this phenomenon, the magnetic moments of the nucleus of the atoms of a sample immersed in a magnetic field become oscillators that rotate around the direction of the field. Resonance phenomena that obey a law qualitatively like that shown in Fig. 1.3(c) result when these oscillators are driven by a sinusoidal electric field. The observation of this resonance provides information useful to study the structure of molecules, which is an issue of interest in many cases.

Other examples could be posed. From Earth sciences: tides are oscillations of the sea surface that result from the periodic gravitational force caused by the Earth's rotation in the presence of the Moon. Or from the engineering of structures, where designs have to be invented to avoid resonances that could result in harmful vibrations. In general, resonance curves like those plotted in Fig. 1.3(c), which are given by equations like Eq. 1.10, describe, at least qualitatively, a widespread phenomenon that results as the response of any linear oscillator that is driven by an external force which varies sinusoidally with time. This is a case frequently found in science and technology and this is why resonance has become a very important phenomenon.

1.2.2 *Coupled harmonic oscillators: normal modes*

It is often found that systems of interest are made of two or more oscillators which interact weakly among them. The dynamics of such systems are both interesting and complex. Some essential results on the dynamics of coupled harmonic oscillators can be obtained by the consideration of two identical free oscillators mutually coupled, which otherwise is a configuration frequently found in practice. The dynamics of this system is given by

$$\frac{d^2x_1}{dt^2} + \omega_0^2 x_1 + \kappa^2 (x_1 - x_2) = 0, \tag{1.12}$$

$$\frac{d^2x_2}{dt^2} + \omega_0^2 x_2 + \kappa^2 (x_2 - x_1) = 0, \tag{1.13}$$

where the two oscillators are labeled 1 and 2, respectively, and κ^2 is a measure of the strength of the coupling. The condition of weak coupling is expressed by the condition $\kappa < \omega_0$. A simple mechanical visualization [Fig. 1.4(a)] can be obtained by considering two identical mass-spring oscillators on a frictionless surface, such as that shown in Fig. 1.1(c), mutually coupled by a third spring.

The time evolution of this system, obtained by the application of the standard techniques for the solution of systems of ordinary linear differential equations, is then given by

$$x_1(t) = A_1 \sin(\omega_1 t + \delta_1) + A_2 \sin(\omega_2 t + \delta_2), \tag{1.14}$$

$$x_2(t) = -A_1 \sin(\omega_1 t + \delta_1) + A_2 \sin(\omega_2 t + \delta_2). \tag{1.15}$$

These solutions are dependent on the frequencies ω_1 and ω_2 which are given by $\omega_1^2 = \omega_0^2 + 2\kappa^2$ and $\omega_2^2 = \omega_0^2$, which can be written in the more symmetric form $\omega_1^2 = \omega_C^2 + \kappa^2$ and $\omega_2^2 = \omega_C^2 - \kappa^2$, when the frequency $\omega_C^2 = \omega_0^2 + \kappa^2$ is introduced. This is the frequency of the harmonic oscillations of any one of the two oscillators, given by Eqs. 1.12–1.13, when the other is held fixed. The constants A_1, A_2, δ_1 and δ_2 are to be determined by the initial values of the variables, $x_1(0)$ and $x_2(0)$, and their rates of change, $\dot{x}_1(0)$ and $\dot{x}_2(0)$.

The main features of the dynamics of two coupled harmonic oscillators are illustrated in Figs. 1.4(b, c). These correspond to particular solutions obtained for initial conditions $\dot{x}_1(0) = \dot{x}_2(0) = 0$, and $x_1(0) = -x_2(0) = A$ [Fig. 1.4(b)], or $x_1(0) = x_2(0) = A$ [Fig. 1.4(c)]. Simple harmonic oscillators, like that presented in Fig. 1.1(a), are obtained in the two cases: the first has a frequency ω_1 and the two oscillators evolve with a phase differ-

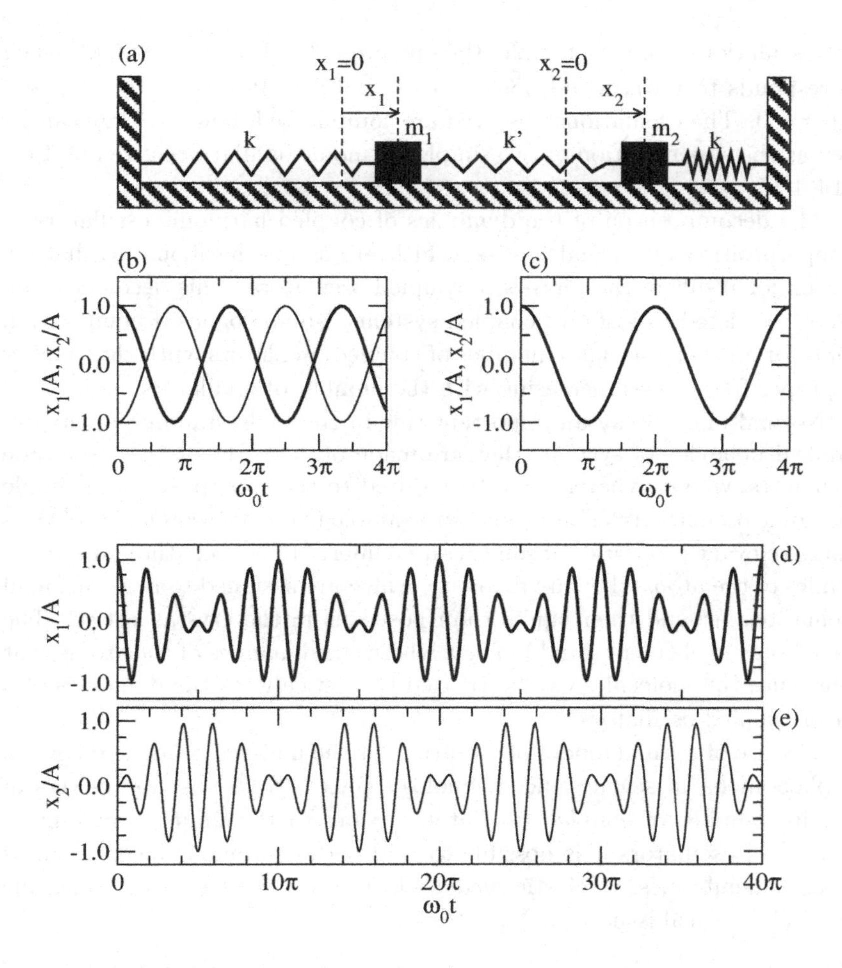

Fig. 1.4 (a) Mechanical model for the system of two coupled harmonic oscillators based on the mass-spring oscillator. Time series for oscillator 1 (thick line) and oscillator 2 (thin line) for (b) the high frequency mode, (c) the low frequency mode, and (d-e) a superposition of the two modes.

ence of $\Delta\phi = \pi$, and the second has a frequency ω_2 and the two oscillators evolve in phase and the two curves superimpose to each other. These are known as the normal modes of vibration of the coupled chaotic oscillators, the first is the mode of high frequency and the second the mode of low frequency. They are interesting because they are simple harmonic oscillations, and any other motions of the coupled systems are superpositions of these two oscillations; as, for example, the one presented in Figs. 1.4(d, e) which corresponds to initial conditions $\dot{x}_1(0) = \dot{x}_2(0) = 0$, and $x_1(0) = A$ and $x_2(0) = 0$. The oscillations observed are complicated, however they can be seen as the superposition of two simple harmonic oscillations given by Eqs. 1.14–1.15.

The decomposition of the dynamics of coupled harmonic oscillators in a superposition of normal modes, which are simple harmonic oscillators, is a major result in the physics of coupled oscillators. This decomposition is not restricted to just two coupled systems: an analogous decomposition holds for an arbitrary large number of coupled oscillators, with the number of proper frequencies increasing with the number of oscillators.

Normal modes play an important role in the understanding of the dynamical behavior of systems that are made of networks of low amplitude oscillators, whose dynamics can be reduced to the superposition of simple harmonic oscillations. This occurs for example in condensed matter physics where relevant properties of solids can be inferred from an study of the dynamics of the atoms they are made up, which are assumed to perform small oscillations around their equilibrium positions in the crystal lattice. The same holds in chemistry and biology, where the dynamics of the atoms that constitute the molecules can be treated theoretically as the dynamics of a set of coupled oscillators.

The use of normal modes of vibration for the understanding and control of phenomena in science and technology goes beyond the case of sets of a finite number of coupled oscillators: going to the limit of an infinite number of oscillators it is possible to deal with continuous media such as strings, membranes, or elastic media which are present in many scientific and technological issues.

1.3 Nonlinear oscillators

All the above discussions on linear oscillators illustrate several important properties of this kind of system: they describe motions around stable equilibria, dissipation causes the fall down to the equilibria and the loss of memory of initial conditions, and external forcing combined with dissipation allows sustained oscillations that are independent of the initial conditions. The last are called limit cycle oscillations.

However, in practice, nonlinearity is very important. This is illustrated in Fig. 1.1(d) which shows that the simple quadratic potential function, which corresponds to harmonic oscillations, only works in the close vicinity of the equilibrium point, which is in fact a small region. Moreover, it is clear that the driven harmonic oscillator (Eq. 1.7) is a particular case of a more general form of dynamical system whose equations of motion read

$$\frac{d^2x}{dt^2} + f\left(x, \frac{dx}{dt}\right) + \frac{dV(x)}{dx} = A_1 \sin(\omega_1 t), \tag{1.16}$$

where $V(x)$ is an arbitrary potential function allowing one stable equilibrium at least, and $f(x, \dot{x})$ is a damping term not necessarily linear. A particular example is the Duffing oscillator whose dynamics are given by

$$\frac{d^2x}{dt^2} + \beta\frac{dx}{dt} - ax + bx^3 = A_1 \sin(\omega_1 t), \tag{1.17}$$

which is a representative case of a linearly dumped oscillator, which moves in a two-well quartic potential, $V(x) = (a/2)x^2 - (b/4)x^4$, subject to a external periodic force. The potential $V(x)$ is enough to make the system nonlinear; i.e. a linear combination of solutions is not a solution anymore.

This model allows an experimental realization by means of a simple magneto-mechanical model [Moon (1987)]. This is made up [Fig. 1.5(a)] of a vertical steel elastic beam fixed at its upper end to a vibrating support which provides the chaos enhancing force $A_1 \sin(\omega_1 t)$. The wire is acted on at the lower free end by two magnets that create two equilibrium points, and then the two-well potential, $V(x)$ [Fig. 1.5(b)]. Eq. 1.16 allows many different significative realizations, which include the Duffing equations, and others such as the equations of the motion of a driven pendulum undergoing large amplitude oscillations

$$\frac{d^2x}{dt^2} + 2\gamma\frac{dx}{dt} + \omega_0^2 \sin(x) = A_1 \sin(\omega_1 t). \tag{1.18}$$

The results summarized in the first paragraph of this section, although illustrated in the above sections with linear oscillators, do also hold for nonlinear oscillators. Nonlinearity, however, has a new major effect on the dynamical behaviors to be observed: nonlinear systems allow new kinds of solutions for the oscillations around the equilibrium which have been named aperiodic solutions. The time series for the system observable, $x(t)$, when an aperiodic solution settles down, oscillates around the equilibrium; however, there is no a definite form of the function $x(t)$, limited to a finite time interval that repeats itself, as in Fig. 1.1(a). Moreover, the phase space plots of such systems do not converge to closed curves like those shown in Fig. 1.1(e) and Fig. 1.3(b), nor fall down to the equilibrium, as in Fig. 1.2(b); instead, they trace entangled trajectories that never repeat themselves.

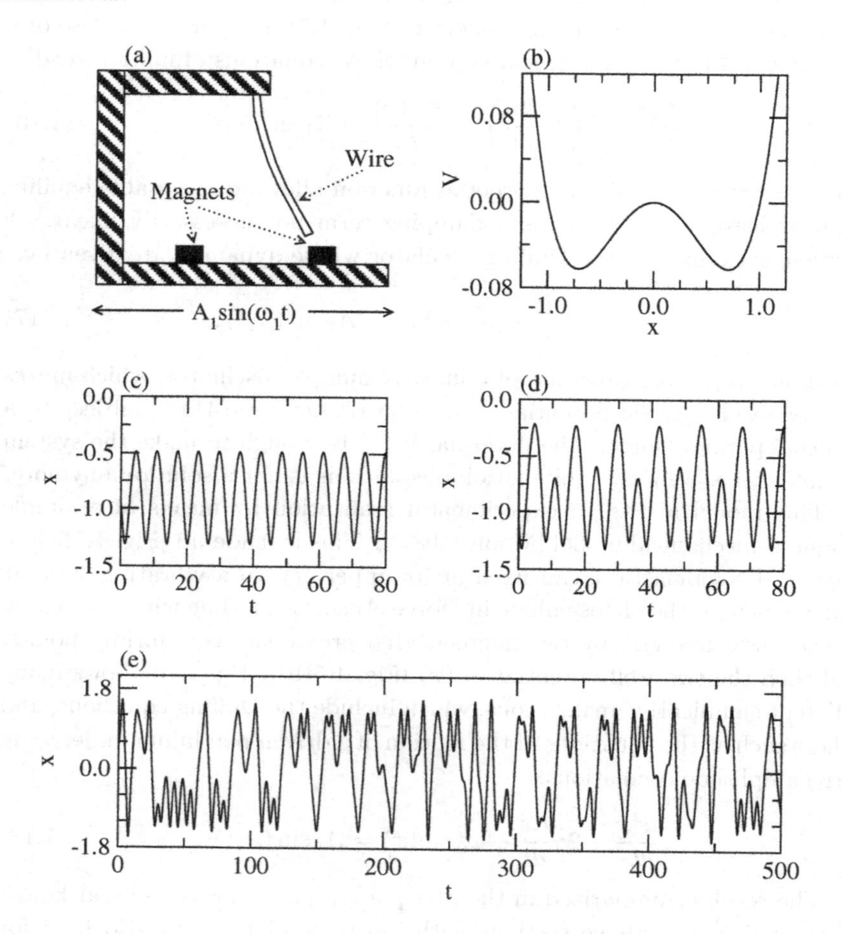

Fig. 1.5 (a) Mechanical model for the Duffing oscillator. (b) The double well potential. Time series for (c) a period one solution ($A_1 = 0.12$), (d) a period four solution ($A_1 = 0.18$), and (e) an aperiodic solution ($A_1 = 0.24$).

Nonlinear systems can be modelled by means of nonlinear differential equations, like Eq. 1.16. There is not a general method to solve nonlinear equations like the one that is employed to obtain the motions of the several types of harmonic oscillators [Riley et al. (1998)] studied above. In general, and especially when aperiodic solutions are involved, the motion of the oscillator in phase space has to be obtained by means of one or more of the numerical methods that have been developed to solve differential equations [Press et al. (1992)]. Then, although there are analytical tools to deal with nonlinear systems [Jordan and Smith (1990)], computational methods play an important role in the study of nonlinear dynamical systems.

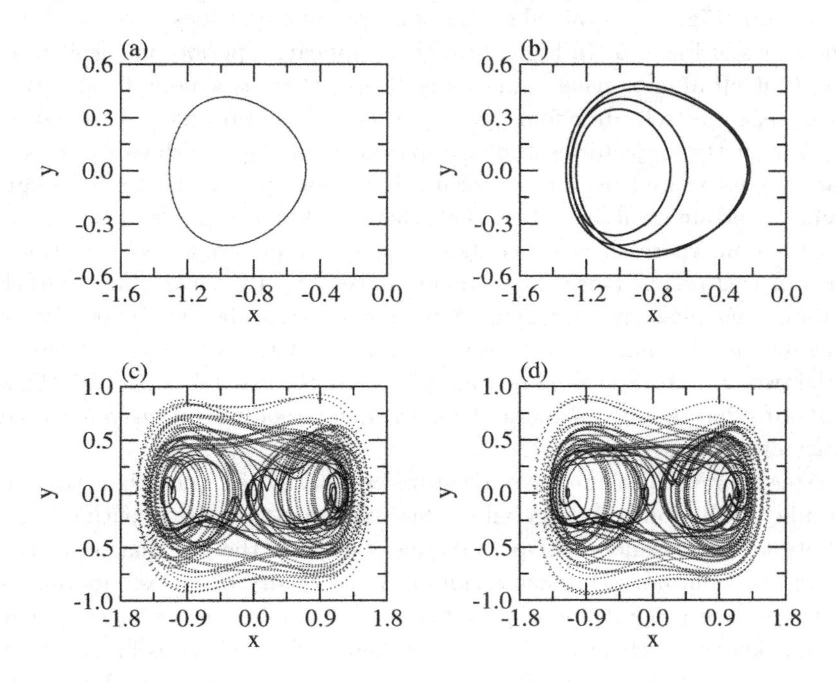

Fig. 1.6 Phase space dynamics for the Duffing oscillator projected onto the $x - y$ plane for: (a) a period one cycle ($A_1 = 0.12$), (b) a period four cycle ($A_1 = 0.18$), and (c,d) an aperiodic attractor ($A_1 = 0.24$) obtained from a trajectory of 10^4 time steps, with the systems started at different initial conditions in each of the two plots.

The Duffing oscillator, Eq. 1.17, can be used to present examples of nonlinear oscillations. Several numerical solutions appear in Figs. 1.5(c-e) for the system parameter values $\beta = 0.10$, $a = b = 1/2$, $\omega_1 = 1$, and A_1 used as a control parameter to select different dynamical behaviors. For small A_1 [Fig. 1.5(c)], the time evolution of the position of the oscillator follows a periodic oscillation around one minimum of the potential function, which looks like that of an harmonic oscillator. An increase in A_1 [Fig. 1.5(d)] makes the oscillation more complex, but this still holds a periodic structure, with a period which is four times that of the previous oscillation. Further increase of A_1 results in an irregular oscillation [Fig. 1.5(e)] which erratically changes the center of oscillation between the two equilibria available. No period can be defined in this case and because of this, this is called an aperiodic oscillation.

Additional illustration of nonlinear dynamics is provided by the phase space plots [Fig. 1.6] obtained for the same parameter values as those of the time series of Fig. 1.5. In Fig. 1.6(a) there appears a period one orbit that repeats itself after a cycle, and in Fig. 1.6(b) there is a period four orbit which repeats itself after four cycles. Aperiodic motion has developed in Fig. 1.6(c): the trajectories in phase space never close and the set of phase space points visited by the representative point appear cloudy, although having a certain structure. This is characterized by the shape of the phase space region where they are contained, and the preference to visit some parts of this region more than others, as seen by the clearer and darker zones in that phase space region. Although an aperiodic orbit started at a different initial condition will result in a different trajectory the structure of the two trajectories would remain essentially the same [Fig. 1.6(d)]. This illustrates that despite its aperiodicity, the non periodic solutions have some invariant structure.

Aperiodic oscillations have a nature essentially different than that of periodic oscillations. A large effort has been made by the scientific community to understand such aperiodic motions since the last quarter of the Twentieth Century. This has resulted in the theory of chaos: aperiodicity is just one property of a special case of dynamics proper of nonlinear systems known as chaotic dynamics. In view of the enormous impact that the results on harmonic oscillations have had in all science and technology, the study of chaotic oscillators is pertinent and relevant because of its intrinsic scientific interest, because of its potential utility for analysis and explanation in all scientific fields, and because of the possible applications in medicine and engineering. In particular, over the last fifteen years, in the subfield of physics known as dynamics of nonlinear systems and chaos, there has been notable activity aimed at the study of coupled and driven chaotic oscillators. The main results of such research have been the discovery of several forms of synchronization, as well as some scenarios of suppression and control of chaos. The present book is aimed to give an account of that research. In the next chapter the main properties of chaotic oscillators will be reviewed, and the remainder of the book will be devoted to the study of the dynamics of coupled and driven chaotic oscillators.

Chapter 2

Chaotic Oscillators

The aim of this chapter is to review basic concepts of the theory of chaos needed in the remainder of the book. The idea behind it is to make the book self-contained. The definition of chaos and the concepts and methods used to characterize and measure it will be introduced and illustrated by means of several chaotic systems. To emphasize the multidisciplinary interest of the subject, these are borrowed from different disciplines including electronics, optics, biology, chemistry, and geophysics. These models will be used later to present specific examples throughout the book.

2.1 Nonlinear flows and maps

Nonlinear continuous flows and discrete maps are two types of mathematical concepts that are used to model the chaotic behavior of many systems frequently found in science and technology. Sometimes the description obtained is a precise and accurate reproduction of what is found in nature or in the laboratory, while in others it just reproduces the main qualitative features of the dynamics. Anyway, they have proven to be very useful tools in the understanding of chaotic dynamics.

Nonlinear oscillators are not limited to systems that allow a modelization by means of the equation of a nonlinear oscillator like Eq. 1.16. In fact, the model for a nonlinear oscillator most commonly found is an autonomous flow. The equations of motion for a general autonomous flow of dimension d can be written as

$$\frac{d\mathbf{x}}{dt} = \mathbf{F}\left(\mathbf{x};\mathbf{p}\right), \tag{2.1}$$

being $\mathbf{x} = (x_1, x_2, ..., x_d)$ the variables that describe the system, and

$\mathbf{F} = [F_1(\mathbf{x};\mathbf{p}), F_2(\mathbf{x};\mathbf{p}), ..., F_d(\mathbf{x};\mathbf{p})]$ a nonlinear vector function, that characterizes the system, and is called the vector field. This depends on a set of parameters $\mathbf{p} = (p_1, p_2, ...)$ which describe the features of a particular realization of the generic systems modelled by the vector field. An autonomous flow is a model for the case when the number of variables needed for a complete description of the system is not necessarily two, but an arbitrary number d. The structure of the system can be quite complex, and is described by the vector field, $\mathbf{F}(\mathbf{x};\mathbf{p})$, which includes the relation between the system variables, $\mathbf{x}$, and the possibility of having systems that have the same structure but are different parametrically.

An important property of a flow is its set of fixed points, $\mathbf{x}_F$. These are given by the condition $\mathbf{F}(\mathbf{x}_F;\mathbf{p}) = 0$, and are phase space points for which $d\mathbf{x}/dt = 0$, so that if the system is in one of these points it will stay there forever unless some perturbation is applied. Fixed points correspond to equilibria, and are said to be unstable when an infinitesimal perturbation applied to the system in the fixed point will result in a trajectory that diverges from the point. Otherwise, fixed points are said to be stable. The dynamics of the flow occurs around the stable fixed points, although unstable fixed points do also play a relevant role. Flows do allow periodic and aperiodic oscillations for the time series of each of the system variables, $\mathbf{x} = (x_1, x_2, ..., x_d)$. It is assumed that, when a system described by a flow is left to evolve from appropriate initial conditions, it will fall towards a set of points, $\mathcal{A} \subset \mathbb{R}^d$, called the attractor, which is bound and dependent on the form of the vector field, $\mathbf{F}$. This can be a fixed point, a limit cycle, a quasi-periodic trajectory, or a chaotic attractor.

It is to be noted that the equations of dynamical systems of the type given by Eq. 1.16 can be rewritten in the form of autonomous flows (Eq. 2.1). This is achieved by means of the introduction of two new variables, y and z, defined as $y = dx/dt$, and $z = \omega_1 t + 2\pi n$, with n a whole number. Then Eq. 1.16 becomes the following three-dimensional autonomous flow

$$\frac{dx}{dt} = y, \tag{2.2}$$

$$\frac{dy}{dt} = -f(x,y) - \frac{dV(x)}{dx} + A_1 \sin(z), \tag{2.3}$$

$$\frac{dz}{dt} = \omega_1. \tag{2.4}$$

When the flow is constructed from Eq. 1.16, this has a significant theoretical structure because it explicitly contains two elements necessary to maintain sustained oscillations: besides the natural oscillation of the system generated by the potential function $V(x)$, it includes explicitly a dissipation term, $f(x, dx/dt)$, and a forcing term, $A_1 \sin(\omega_1 t)$, whose competition is needed for that aim. For a generic flow these elements may not appear so explicitly, but they will be implied in some form.

An example of three-dimensional flow which allows chaotic dynamics is the Lorenz model [Lorenz (1963)]. This has an historical relevance because the study of its chaotic solutions has played an important role in triggering the interest towards chaotic dynamics. It has received a lot of attention in the literature, and is frequently used to demonstrate properties of chaotic systems. The Lorenz model is aimed to be a very simple model which catches the essence of the dynamics of a fluid heated from below in the gravitational field. Because this is the physical situation that occurs on the Earth's, and other planetary atmospheres, it was first used to study the problem of the weather predictability [Lorenz (1963)]. The $d = 3$ flow that describes this system is

$$dx/dt = \sigma(y - x), \tag{2.5}$$

$$dy/dt = x(r - z) - y, \tag{2.6}$$

$$dz/dt = xy - bz, \tag{2.7}$$

being $\mathbf{x} = (x, y, z)$ the variables that define the state of the system, and $\mathbf{p} = (\sigma, r, b)$ the set of parameters that characterize the particular properties of the fluid and its geometrical conditions. For the parameter values $\sigma = 10$, $r = 28$, and $b = 8/3$, the vector field $\mathbf{F} = [10(y - x), x(28 - z) - y, xy - 8/3z]$ leads to an attractor $\mathcal{A} \subset \mathbb{R}^3$ which is chaotic.

The experimental situation modelled by the Lorenz equations is presented in Fig. 2.1(a): a fluid sample in the gravitational field, g, interacts with two heat sources, one at high temperature, T_H, and the other at low temperature, T_L. The buoyancy force caused by the temperature difference, and the weight due to gravity compete to cause a rotational movement of the fluid know as convection (circles). Moreover there is dissipation caused by viscosity. The system variables measure: the intensity of the convective motion, x, the temperature difference between ascending and descending elements, y, and the deviation of the vertical temperature profile from linearity, z [Lorenz (1963)]. The Lorenz model for the parameters said above provides an example of turbulent convection. This is illustrated by means of the aperiodic nature of the trajectory followed by the system in the

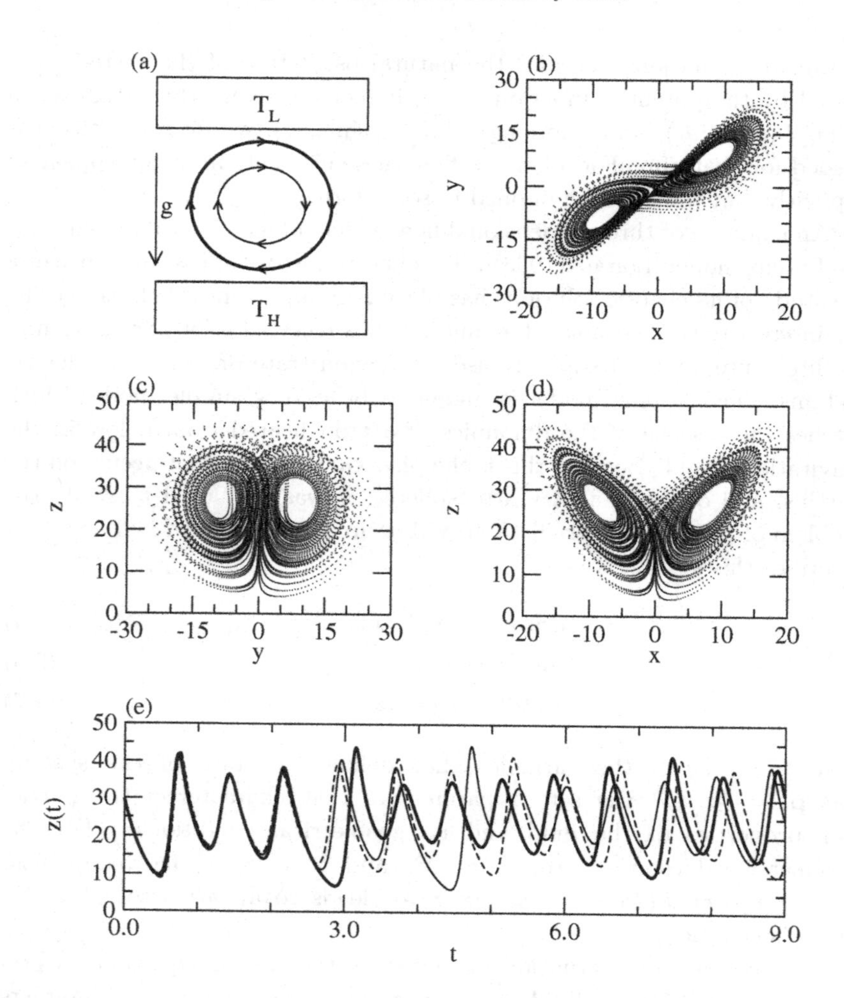

Fig. 2.1 (a) Schematic view of the elements involved in the convective fluid dynamics described by the Lorenz equation: heat sources (rectangles), gravity force (arrow), and convective rolls (arrowed circles). (b–d) Chaotic attractor of the Lorenz model projected onto three orthogonal planes. (e) Aperiodic oscillations of $z(t)$ for three time series started at different but very close initial conditions.

three-dimensional phase space, (x, y, z). This trajectory, computed for a sufficiently large time interval, provides an image of the chaotic attractor, $\mathcal{A} \subset \mathbb{R}^3$, as shown in Figs. 2.1(b-d) which presents 10^4 points for a trajectory obtained numerically. The complex structure observed is a signature of chaos. However, strongest evidence of the chaotic dynamics in

$\mathcal{A}$ is provided by the observation of the time evolution of different trajectories started at very close points in phase space. This is illustrated in Fig. 2.1(e), where three time series for $z(t)$ started at initial conditions, (x_0, y_0, z_0), whose difference was of about 1 per cent are presented. Up to $t \approx 2.6$, the three time series are practically identical; however for times larger than 2.6 they evolve independently. This means that for an initial condition, (x_0, y_0, z_0), measured with an error of 1 per cent, predictability of its future state becomes lost at $t \approx 2.6$, after approximately 3.0 cycles. This loss of predictability is the essential feature of chaos.

There is a special and interesting case of potentially chaotic systems whose dynamics occur in a continuous time and are given by differential equations, but that cannot be described as autonomous flows. These are systems described by means of delayed equations

$$\frac{dx}{dt} = F(x, x_\tau), \tag{2.8}$$

with $x = x(t)$ and $x_\tau = x(t - \tau)$ the values of the state variable at times t and $t - \tau$ respectively. The variable x_τ is known as the delayed variable, and τ is the time delay. An example is the Mackey and Glass model for physiological control [Mackey and Glass (1977)]

$$\frac{dx}{dt} = -\alpha x + \frac{\beta x_\tau}{1 + x_\tau^C}, \tag{2.9}$$

with α, β, and C characteristic parameters. Another example is given by the Ikeda equation

$$f(x_\tau) = -\alpha x - \beta \sin(x_\tau) \tag{2.10}$$

[Ikeda et al. (1980)] which is used in nonlinear optics. These kinds of systems are said to evolve in an infinite-dimensional phase space [Farmer (1982)], because initial conditions in a continuous time interval of length τ are needed to define a particular solution.

Besides continuous flows and delayed equations, other kinds of dynamical systems used very often are the discrete maps [Elaydi (2000)]. When dealing with them, the dynamical variables $\mathbf{x} = (x_1, x_2, ..., x_d)$ evolve in a series of consecutive steps numbered by integer numbers, $j = 0, 1, 2, ...$, instead of in a continuos time, t, specified by real numbers. The equations of motion are then given by a rule of the type

$$\mathbf{x}_{j+1} = \mathbf{F}(\mathbf{x}_j), \tag{2.11}$$

with $\mathbf{F} = [F_1(\mathbf{x}), F_2(\mathbf{x}), ..., F_d(\mathbf{x})]$ a vector function that defines the map. The system evolves in time, from certain initial condition, $\mathbf{x}(0)$, by means of successive iterations of Eq. 2.11, which is called a difference equation, to produce a discrete series of points $\mathbf{x}(1)$, $\mathbf{x}(2)$, ... which define the trajectory of the system in phase space. For the systems of interest, as before, this trajectory occurs in an attractor $\mathcal{A} \subset \mathbb{R}^d$ which may be of one of several types, including chaotic attractors.

An example of a two-dimensional map, frequently found in the literature is the Hénon map [Hénon (1976)],

$$x_{j+1} = p - a \cdot x_j^2 + y_j, \tag{2.12}$$

$$y_{j+1} = b \cdot x_j, \tag{2.13}$$

whose behavior is given by the system parameters p, a and b. This is a model invented with a mathematical motivation to investigate the properties of chaos. In particular, it will be used here to illustrate the complex structure that is characteristic of chaotic attractors. For the parameter values $p = 1$, $a = 1.4$, and $b = 0.3$ the dynamics of this system occurs in a chaotic attractor [Hénon (1976)]. An image of this attractor is presented in Fig. 2.2(a): it appears to have a simple structure made up of a folded curve; however, when particular regions of this curve are enlarged new details are unveiled. This is illustrated in Fig. 2.2(b) that shows an enlargement by an order of magnitude of the top-left tip of the curve, which appears to be made of two curves. More enlargement of the tip of the outer curve shows six new curves [Fig. 2.2(c)], which after further enlargement still display newer details [Fig. 2.2(d)]. The existence of details for ever increasing magnification is a signal of what has been called a fractal structure [1]. This is characteristic of chaotic attractors, which are said to be strange attractors because they present this property.

A very simple, but interesting model of discrete map, is given by the logistic equation

$$x_{j+1} = r \cdot x_j \cdot (1 - x_j), \tag{2.14}$$

which is a one-dimensional system, $d = 1$. This model is of interest to biology and the social sciences for studying population dynamics [May (1976)]. Its single variable, x_j, measures a population, whose rate of growth when the population is very small is given by the system parameter, r. As the population grow a resistance to that increase develops, caused by limited resources for example, which is modelled by the term $(1 - x_j)$. Beyond, its

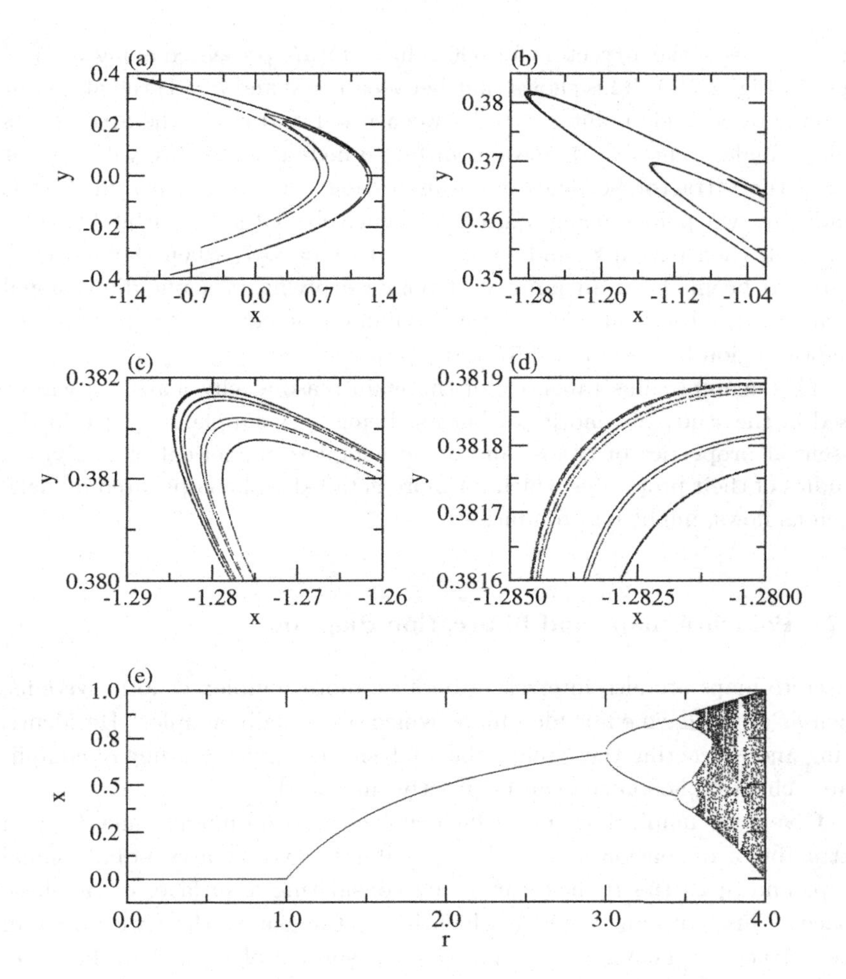

Fig. 2.2 (a) Image of the Henon attractor, and (b, c, d) progressive enlargements of the top-left tip by an order of magnitude each. In all cases 5000 phase space points are used to visualize the attractor. (e) The different types of attractors allowed to the logistic map as the system parameter r is varied.

interest for applications, this model is important in the theory of chaotic dynamics because, being extremely simple, it displays many fundamental features of chaos.

The set of attractors available to this system is displayed in Fig. 2.2(e). Because the dynamics occur in one dimension, a state of the system is represented by a point in a line, and an attractor by a set of points in this line. Images of the attractors for 600 values of r are presented along vertical lines in Fig. 2.2(e). This shows that between $r = 0$ and $r = 1$, the attractor is made by a single point $x = 0$, between $r = 1$ and $r = 3$ the attractor is still a single point $x > 0$ whose coordinate increases steadily with r. For $r > 3$ the attractor becomes more interesting: first it is a period 2 orbit made by two points being visited alternatively, later the orbit becomes period 4, then period 8, and so forth up to $r \approx 3.57$, when the attractor starts to be made of an infinity of points evolving in a one-dimensional strange attractor which is chaotic. Windows of periodicity disturb this chaotic region between $r \approx 3.57$ and $r = 4$.

This example illustrates one of the main reasons why maps are widely used in the study of chaotic oscillators: being very simple, they retain the essential properties of chaos, and allow complete numerical or analytical studies of their properties which for more detailed and complicated models, such as flows, might be prohibitive.

2.2 Poincaré maps and bifurcation diagrams

Discrete maps are also interesting because more complex chaotic systems, such as flows, have embedded maps which are usually simpler. By identifying and extracting these maps the analysis and understanding of complicated chaotic phenomena can be greatly simplified.

Given a dynamical system, which evolves in a continuous time, t, in an attractor of dimension $d > 2$, it is possible to extract a two-dimensional map, known as the Poincaré map, by considering a surface, σ, in phase space. This is usually a plane which is not tangent to the trajectories in the attractor. This surface, know as the surface of section or Poincaré surface, is crossed again and again by the trajectory of the representative point. To define the Poincaré map, one of the two sides of the surface is selected, and the continuous time is substituted by a discrete time by counting the successive times, t_j, when the trajectory intersects the surface from the selected side. Then the coordinates of the intersection point measured on the surface, say $\mathbf{x}_j = [x(t_j), y(t_j)]$, are assigned to the value of j that corresponds to the time of this intersection. Because the value, $\mathbf{x}_{j+1}$, which follows to the value of $\mathbf{x}_j$ is unambiguously defined, a discrete map, $\mathbf{x}_{j+1} = \mathbf{F}(\mathbf{x}_j)$, whose dynamics occur in a two-dimensional phase space is determined. This procedure to create a map from a flow is sketched in Fig. 2.3. The continuous trajectories intersect the plane, σ, at times $t_j, t_{j+1}, \ldots$

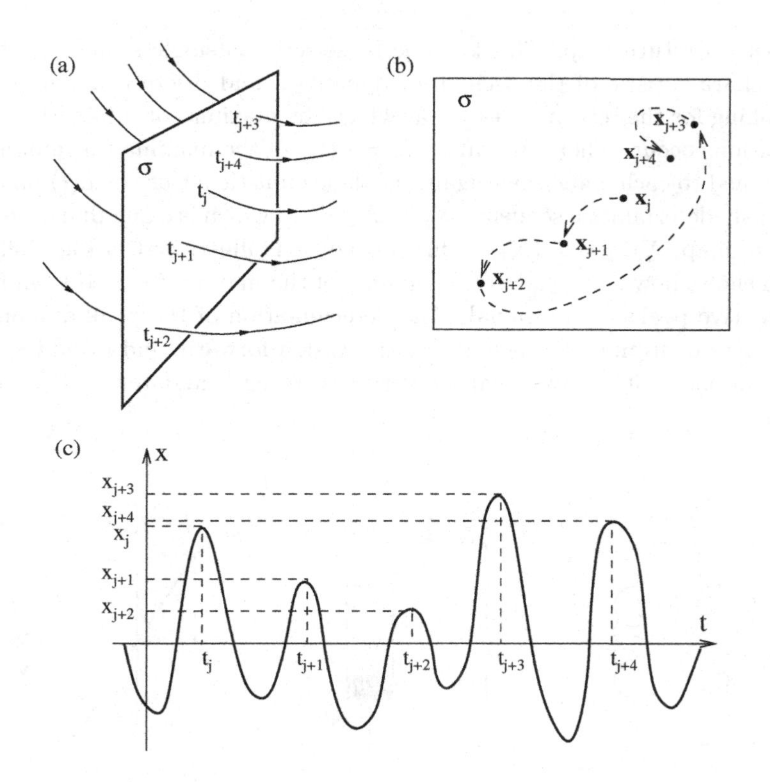

Fig. 2.3 (a) A surface of section, σ, is crossed by a trajectory of a dynamic system whose representative point in phase space moves in an attractor, (b) a Poincaré map, $\mathbf{x}_{j+1} = F(\mathbf{x}_j)$, is defined by the consecutive series of points, $\mathbf{x}_j$ measured on σ, in which the trajectory cuts the surface. (c) A return map, $x_{j+1} = F(x_j)$, is defined by means of the consecutive maxima reached by the scalar variable $x(t)$.

[Fig. 2.3(a)], which are called return times. The sequence of points on σ where these intersections occur, $\mathbf{x}_j$, is completely deterministic, because of the deterministic nature of the trajectory in the whole phase space, so a well defined sequence $\mathbf{x}_j \to \mathbf{x}_{j+1} \to ...$ results, which is the map on σ [Fig. 2.3(b)] called the Poincaré map . This can be obtained numerically from continuous flows, by a numerical integration of the differential equations to find the trajectory, plus an algorithm to detect its intersection with the surface of section. Several useful and simple proposals have been made to this aim [Hénon (1982)].

One-dimensional discrete maps can be obtained from the dynamics of chaotic oscillators that occur in attractors of arbitrary dimension by constructing a return map. This is made by selecting a single scalar variable, $x(t)$, characteristic of the oscillatory dynamics, and discretizing the time by looking for the return times, t_j, at which the maxima (or minima) of the oscillations occur. Then, the value $X_j = x(t_j)$ of the maxima (or minima), is assigned to each j and once again the deterministic nature of $x(t)$ unambiguously determines a sequence $X_1 \rightarrow X_2 \rightarrow \ldots$, which is a one-dimensional discrete map, $X_{j+1} = F(X_j)$. This procedure is illustrated in Fig. 2.3(c), which shows how a sequence of five points of the map is obtained from five consecutive peaks of the signal. The determination of the peak dynamics from flows or from experimental data is straightforward, and given its relative simplicity its analysis can be very rewarding [Candaten and Rinaldi (2000)].

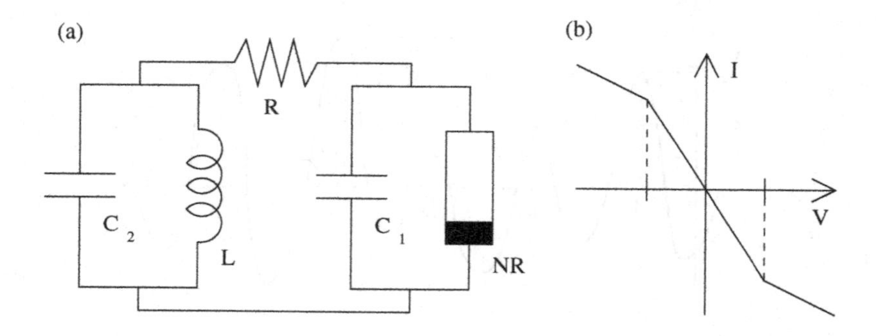

Fig. 2.4 (a) Scheme of the circuit of Chua which is made up of five main components: two capacitors (C_1 and C_2), an inductance (L), a linear resistor (R), and a nonlinear resistor (NR). This last has a piecewise characteristic which is sketched in (b).

The reduction of the dynamics of a flow to these two types of discrete maps will be illustrated by means of the circuit of Chua [Matsumoto et al. (1985)]. This is made up of two sub-circuits: a capacitor connected in parallel with a nonlinear resistor [Kennedy (1992)], and a second capacitor connected in parallel with an inductance; these are mutually connected by means of a resistor (Fig. 2.4). The differential equations that describe the circuit dynamics are [Matsumoto et al. (1985)]

$$dx/dt = \alpha\,[y - x - f(x)], \tag{2.15}$$

$$dy/dt = x - y + z, \tag{2.16}$$

$$dz/dt = -\beta y, \tag{2.17}$$

with

$$f(x) = bx + \frac{1}{2}(a - b)[|x + 1| - |x - 1|], \tag{2.18}$$

where x and y are the voltages in the capacitors 1 and 2 respectively, and z is the current in the inductance. Depending on the choice of parameters (α, β, a, b) a variety of chaotic and non-chaotic attractors are available to this system [Chua et al. (1993)]. This, combined with its simple structure, has made the circuit of Chua a system often used to demonstrate experimentally, theoretically and numerically many properties and phenomena proper of chaotic oscillators.

An example of the reduction of the three-dimensional flow to a two-dimensional map by means of a surface of section is presented in Fig. 2.5. Two attractors corresponding each to a different set of system parameters have been considered: a torus ($\alpha = 1800.0$, $\beta = 10000$, $a = -1.026$, and $b = -0.982$) and a chaotic attractor ($\alpha = -6.69191$, $\beta = -1.52161$, $a = -1.142857$, and $b = -0.714286$). These have been obtained here by numerical integration of Eqs. 2.15-2.18; however, they have also been realized in the laboratory [Chua et al. (1993)]. The motion in a torus [Fig. 2.5(a)] is a special case of non-chaotic motion in which the system dynamics is characterized by two incommensurate frequencies; i.e. two frequencies whose ratio is an irrational number. The trajectories then occur on a surface which has the topology of a donut (a torus): one frequency characterizes a rotation in phase space along a closed curve which is the edge of the torus, while the other corresponds to a rotation around this edge on the surface of the torus. A projection of a three-dimensional trajectory in the particular chaotic attractor considered in this example is presented in Fig. 2.5(b), and displays a set of points with a structure similar to other chaotic sets shown before. Surfaces of section given by $x = x_P$, with x_P a constant equal to the average value of $x(t)$ have been used to obtain the Poincaré maps. These surfaces of section are planes perpendicular of the plane $x - y$, whose intersections with it are straight lines, which are plotted in Figs. 2.5(a, b). The points of the map were chosen when the trajectories intersected the surface in the direction of increasing x. The resulting map for the torus attractor is displayed in Fig. 2.5(c), and shows a closed curve that depicts the intersection of the donut surface with the Poincaré plane. The map for the chaotic attractor appears in Fig. 2.5(d) and bears certain resemblance with the chaotic attractor of the Hénon map. The different dynamical nature between the quasiperiodic and the chaotic attractors is

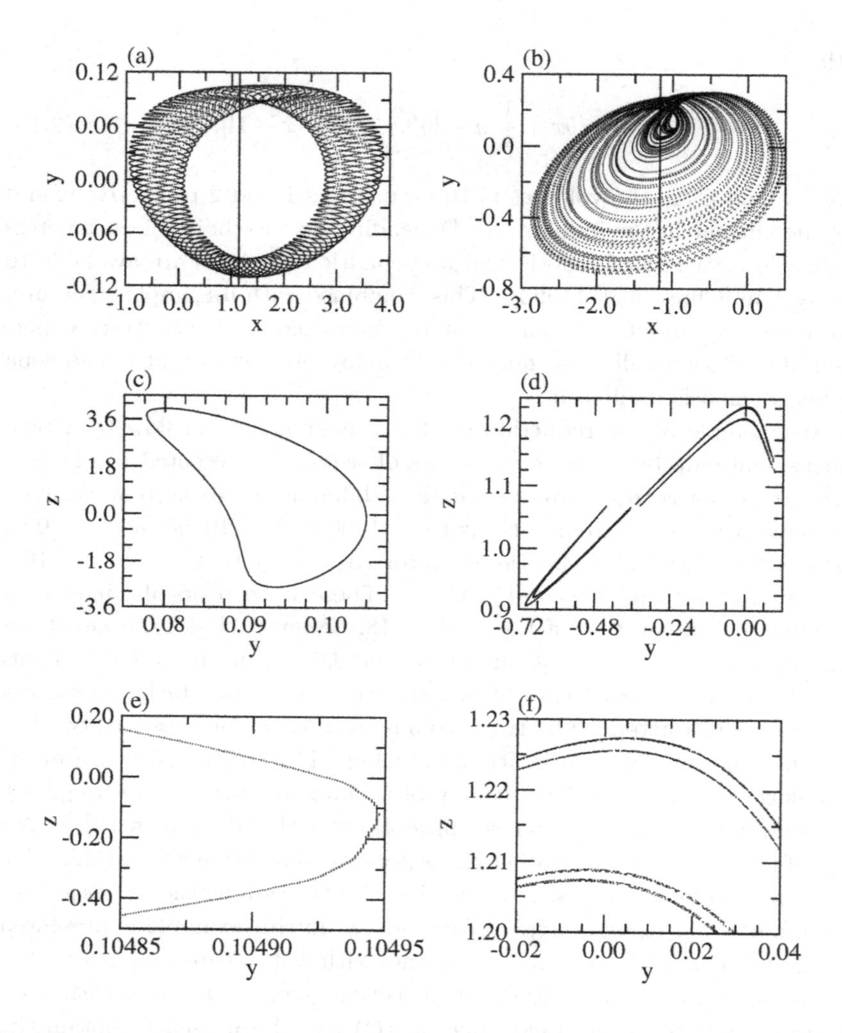

Fig. 2.5 Projection onto the $x-y$ plane of the trajectories of the circuit of Chua tuned to parameter values such that the attractor is: (a) a torus, and (b) chaotic. The straight lines are the intersections of the surfaces of section with the plane $x-y$. The Poincaré maps appear in (c) and (d), respectively, and enlargements of particular regions of these maps in: (e) torus, and (f) chaotic attractor.

inferred from the different structure of the maps. This is more notorious when enlargements of small regions of each map are visualized. The right tip of the non-chaotic torus attractor [Fig. 2.5(e)] displays no further detail; while an enlargement of the top of the map from the chaotic attractor

[Fig. 2.5(f)] shows the emergence of new details as it happened with Hénon map studied above. This example therefore, shows how the structure of the Poincaré map can be useful to distinguish chaotic from non-chaotic motions. A non-trivial case of non-chaotic motion has been considered; if a limit cycle attractor was used, the distinction would have been easier. This is because a periodic orbit, which repeats itself periodically, cuts the surface of section in a well defined finite number of points; therefore, the plot of the map in the Poincaré section is a discrete set made usually of few points.

For the example of the reduction to a one-dimensional map, three-dimensional attractors obtained form Eqs. 2.15–2.18 with $\beta = 14\ 2/7$, $a = -8/7$, and $b = -5/7$, and several values of α have been considered. A particular determination of a return map obtained from the maxima of the $z(t)$, Z_j, for $\alpha = 8.65$ is presented in Fig. 2.6(a). There all the pairs (Z_j, Z_{j+1}) are presented as points in a set of Cartesian axis with Z_j as the abscissa and Z_{j+1} as the ordinate. The resulting plot is a graphical representation of the one-dimensional map $Z_{j+1} = F(Z_j)$. The particular form of F, suggested by Fig. 2.6(b), is that of a parabola; i.e. a quadratic map like the logistic map. A study of the map obtained in this way for 300 values of α in the interval $8.1 \leq \alpha \leq 8.7$ is presented in Fig. 2.6(b); there, all the points Z_j obtained from $z(t)$ are displayed for each value of the system parameter. The resulting figure reproduces the qualitative features of the logistic map presented in Fig. 2.2(e). For small α the map is a single fixed point. With increasing α, this experiences successive bifurcations to period two, period four, period eight attractors, and so on, until a one-dimensional chaotic attractor made of an infinity of points emerges. Some of the correspondent trajectories in phase are illustrated in Figs. 2.6(c-f): a plain periodic limit cycle for $\alpha = 8.15$ [Fig. 2.6(c)], a period two cycle for $\alpha = 8.30$ [Fig. 2.6(d)], a period four cycle for $\alpha = 8.42$ [Fig. 2.6(e)], and a chaotic attractor for $\alpha = 8.50$ [Fig. 2.6(f)]. The strong resemblance between Fig. 2.6(b) and Fig. 2.2(e) deserves to be stressed. This shows how the study of a one-dimensional map embedded in a nonlinear oscillator can easily provide the nature of the dynamical behaviors available to the system as the system parameters are varied.

The particular structures shown in Fig. 2.2(e) and Fig. 2.6(b) provide an example of what has been called the period doubling route to chaos. This is a form of transition from periodicity to chaos which is universal in the sense that its fundamental features [Feigenbaum (1978); Feigenbaum (1983)] are displayed by many mathematical and real systems. Several other

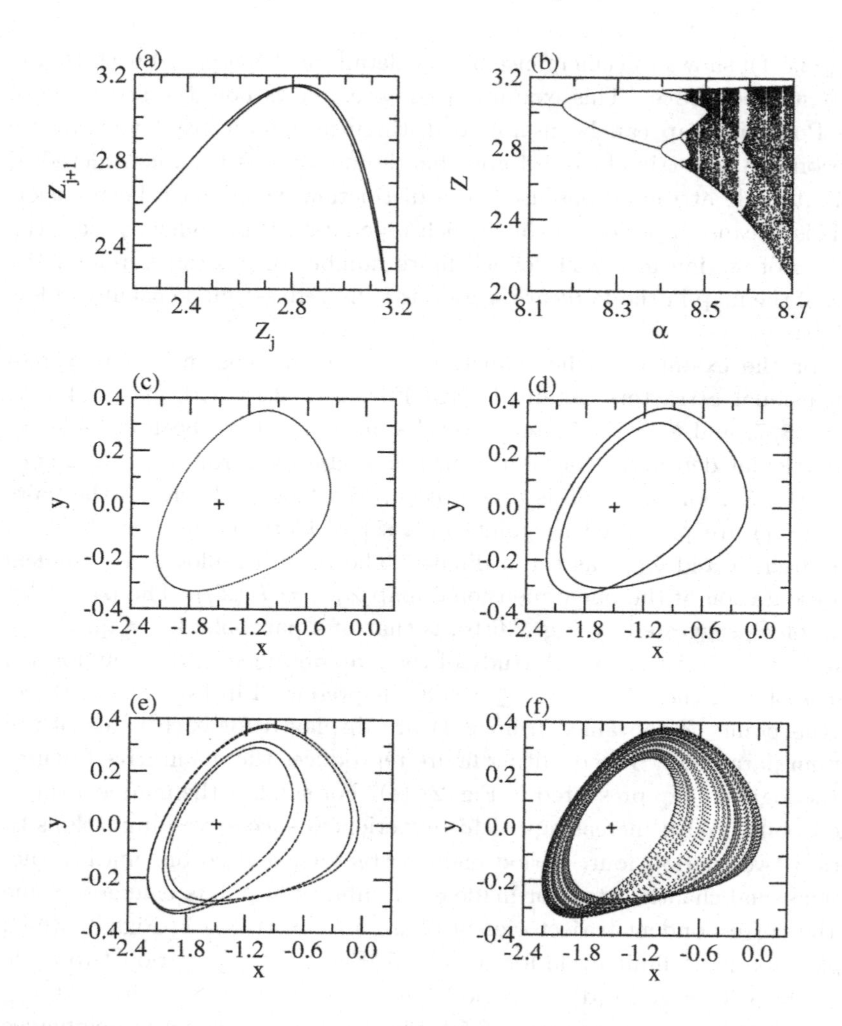

Fig. 2.6 (a) Plot of the return map obtained from the maxima of $z(t)$ for the circuit of Chua for $\alpha = 8.65$. (b) Bifurcation diagram obtained from $z(t)$ when α is varied in the range $[8.1, 8.7]$. Projection of the system trajectories onto the plane $x - y$ for increasing values of a system parameter: (c) $\alpha = 8.15$, (d) $\alpha = 8.30$, (e) $\alpha = 8.42$, and (f) $\alpha = 8.50$. A stable fixed point is indicated by a cross.

routes to chaos have been observed [Eckmann (1981); Swinney (1983)]. Together with the period doubling route to chaos, one of the first to be discovered is the transition to chaos through quasi-periodicity: a system evolving in a limit cycle, changes to a motion in a torus, and then to a chaotic attractor when a system parameter is modified [Ruelle and Takens (1971); Newhouse et al. (1978)]. A whole part of the theory of dynamical systems and chaos, know as bifurcation theory, has been developed to deal with the situations in which a change in the parameters of a system causes a qualitative change in the nature of its dynamical behavior. Several kinds of transitions from periodicity to chaos, besides those mentioned above, and precise characterizations of them are provided by bifurcation theory [Guckenheimer and Holmes (1983); Crawford (1991)]. Besides the routes to chaos, other kinds of transitions are of interest, in particular chaos–chaos transitions are important when chaotic oscillators are studied. An example of such transitions is the interior crisis which is a sudden change of the attractor size when a system parameter is modified. Most of the interior crisis studied until now are discontinuous [Grebogi et al. (1982)]; however, recently the existence of continuous crisis has been observed [González-Miranda (2003)].

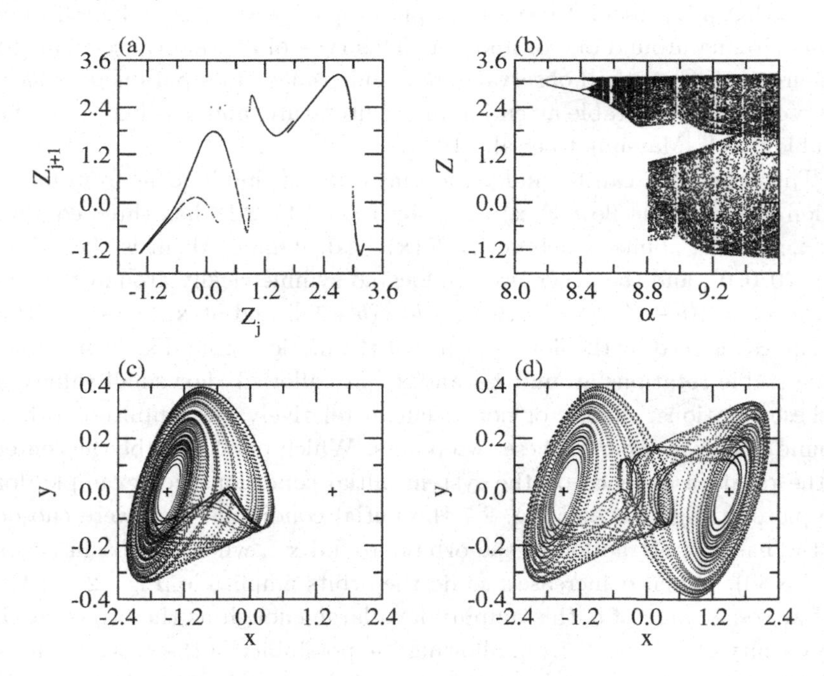

Fig. 2.7 (a) Plot of the return map obtained from the maxima of $z(t)$ for the circuit of Chua for $\alpha = 9.40$. (b) Bifurcation diagram obtained from $z(t)$ when α is varied in the range $[8.0, 9.6]$. Projection of the system trajectories onto the plane $x - y$ for : (c) $\alpha = 8.80$, and (d) $\alpha = 8.84$. Stable fixed points are indicated by crosses.

For additional illustration of maps and bifurcation diagrams an example of chaos–chaos transition on the circuit of Chua is presented in Fig. 2.7. This continues the previous example illustrating the period doubling route to chaos (Fig. 2.6) and the same parameter values have been used. The plot of a return map for $\alpha = 9.40$, presented in Fig. 2.7(a), shows a more complicated relation for $Z_{j+1} = F(Z_j)$, than the map in Fig. 2.6(a), which suggests that for larger values of α, in the chaotic regime, the system dynamics have to become more complex. An expanded bifurcation diagram, $8.0 \leq \alpha \leq 9.6$, shows a sudden change in the range of values reached by Z_j when $\alpha > 8.8$, which is a signal of a substantial change in the chaotic dynamics. This is illustrated in Figs. 2.7(c, d) by means of plots of the projection of the system trajectory onto the $x - y$ plane for $\alpha = 8.80$, and for $\alpha = 8.84$, respectively. This parameter change, smaller than 0.5 percent, changes the topology of the chaotic attractor from a chaotic orbit rotating around a single center, to a chaotic orbit, which has two centers of rotation so that a representative point in phase space goes back and forth from orbits circling around one or another. This type of attractor, accessible to the circuit of Chua, and observable in a wide range of its parameter space, has received considerable attention in the literature, and it is known as the Double Scroll [Matsumoto et al. (1985)].

This transition can be understood in terms of simple ideas from bifurcation theory. The flow $\mathbf{F}(\mathbf{x})$ given by Eqs. 2.15–2.18 has three equilibria; i.e. three points which verify $\mathbf{F}(\mathbf{x}) = \mathbf{0}$. One of them is the origin $\mathbf{x}_0 = (0, 0, 0)$, and the other two are located symmetrically around the origin at $\mathbf{x}_+ = [(b-a)/(b+1), 0, (a-b)/(b+1)]$, and at $\mathbf{x}_- = -\mathbf{x}_+$. The dynamics induced by the flow is such that the motion around $\mathbf{x}_0$ is unstable, while stable rotations around $\mathbf{x}_+$ and $\mathbf{x}_-$ are allowed. For small values of α these rotations, chaotic or not, occur in relatively low amplitude orbits around one or another of these two points. Which point is to be the center of the rotation depends on the system initial conditions; for example, for the plots in Fig. 2.6 and Fig. 2.7 the initial conditions used were chosen on the basin of attraction to the orbits around $\mathbf{x}_-$, when this basin exists ($\alpha \lesssim 8.80$). When α increases, so do the orbits amplitude [Figs. 2.6(c–f)], and at some value of α this amplitude is large enough as the orbits reach the vicinity of $\mathbf{x}_0$ [Fig. 2.7(c)], allowing the possibility of the representative point to be thrown around the opposite equilibria [Fig. 2.7(d)]. A chaotic attractor with a Double Scroll structure then results from an irregular series of back and forth jumps between rotations around $\mathbf{x}_+$ and $\mathbf{x}_-$, that occur through $\mathbf{x}_0$ [Chua et al. (1992)].

2.3 Lyapunov exponents

The essential property of chaos is an intrinsic instability of chaotic trajectories that causes unpredictability of the future state of the chaotic system. This unpredictability is called sensitivity to initial conditions because two trajectories of the same system, that start in states so close that they are indistinguishable within experimental uncertainties, would diverge exponentially. Therefore, for times greater than a certain time characteristic of the system, T_p, the numerical value of their distance in phase space will fluctuate around a quantity whose size has the same order of magnitude as the attractor size. This means that after T_p, predictability has been completely lost: the state of the system may be any of all the available.

A precise and complete characterization of the stability properties of the orbits of a dynamical system is provided by its spectrum of Lyapunov exponents [Benettin et al. (1980a)]. For a system whose dynamics occurs in a phase space of dimension d, this is a set of d real numbers, $\{\lambda_1, \lambda_2, ..., \lambda_d\}$, ordered from largest to smallest, which measure the average rate of divergence, from an orbit of the attractor, of other close orbits started in points which lie along d orthogonal principal directions.

An idea of the meaning of the Lyapunov exponents can be obtained [Wolf et al. (1985)] by choosing an arbitrary point in its stable attractor, $\mathbf{x}(0)$, and considering an infinitesimal hypersphere, $S_d(0)$, of radius R in the d-dimensional phase space centered around this point. All the points inside this hypersphere represent possible initial conditions in the close neighborhood of $\mathbf{x}(0)$. Then, the time evolution of $\mathbf{x}(0)$, together with the points inside $S_d(0)$, under the equations of motion is followed for a short time. This will result in a reference trajectory in the attractor, $\mathbf{x}(t)$, from $\mathbf{x}(0)$; while the evolution of the points in $S_d(0)$, will turn this hypersphere to the shape of an hyperellipsoid, $E_d(t)$, centered in $\mathbf{x}(t)$. This is because the rate of divergence of the trajectories that start in the points initially in $S_d(0)$ will be different along different directions. Each of the axes of the hyperellipsoid provides a principal direction, and the time variation of the length of this axis, $L_i(t)$, from its initial value, R, provides the numerical value of the associated Lyapunov exponent by means of

$$\lambda_i = \lim_{t\to\infty}\left[\frac{1}{t}\ln\left(\frac{L_i(t)}{R}\right)\right]. \tag{2.19}$$

When the values of $L_i(t)$ stays around R, λ_i will be a null exponent. An exponential increase of $L_i(t)$ provides a positive exponent, and an exponential decrease of $L_i(t)$ a negative exponent. Because there are as many Lyapunov exponents as the dimension of phase space, each chaotic system has an spectrum of Lyapunov exponents, $\lambda_1 \geq \lambda_2 \geq ... \geq \lambda_d$. The direction of the axes of the ellipsoid $E_d(t)$ change with time, therefore there is no well-defined direction associated to each Lyapunov exponent. Moreover, because infinitesimal hyperspheres and hyperellipsoids have been considered, Lyapunov exponents are local quantities in the sense that they do not describe the dynamics of orbits that have a large separation.

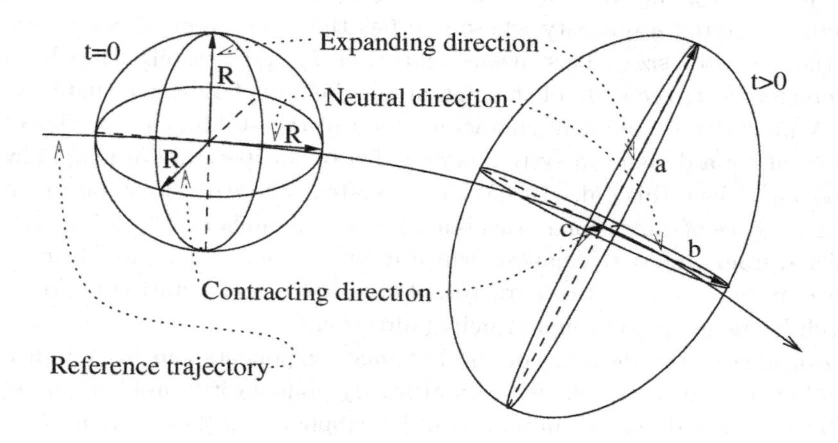

Fig. 2.8 Stability properties of a trajectory, $x(t)$, in a three dimensional chaotic attractor. At $t = 0$ an infinitesimal sphere of initial conditions, of radius R, centered in $x(t)$ starts to evolve under the dynamic equations (Eq. 2.1). At $t > 0$ the radius along the expanding direction has growth to a value $a > R$, and along the contracting direction it has shrunk to a value $c < R$. Along the neutral direction no significant changes have occurred ($b \approx R$); therefore, the sphere has become an ellipsoid of axes $a > b > c$.

The spectrum of Lyapunov exponents of a given system characterizes the nature of its dynamics [Shimada and Nagashima (1979); Benettin et al. (1980a)]. These can be understood from the fact that for each exponent, λ_i, $i = 1, 2, ..., d$, at each point in the trajectory there is a direction along which the projection of the vector position of a close trajectory changes as $d_i(t) \sim d_i(0) \cdot \exp(\lambda_i t)$. Then, if there is at least one positive Lyapunov exponent, there is exponential divergence of trajectories, and the system is chaotic; so the knowledge of the sign of the first Lyapunov exponent is enough to establish if a system is chaotic or not. If the first Lyapunov exponent is negative, all other exponents will be negative, and the stable dynamics of the system is a fixed point. This is because, in this case, all close trajectories will collapse on $\mathbf{x}(t)$; in particular, initial conditions chosen as points of the curve $\mathbf{x}(t)$, advanced or delayed respect to $\mathbf{x}(0)$, will follow trajectories that tend to collapse on a single point. In order to have trajectories, chaotic or not, that do not collapse to a point there must be a zero Lyapunov exponent [Haken (1983)], which corresponds to a direction along which there is no convergence nor divergence of close trajectories. This direction has to be tangent to each point in the trajectory.

Another property of the Lyapunov exponents worth noting is that the sum of all Lyapunov exponents gives the change of elementary volumes, such as the above ellipsoid, around the attractor:

$$V(t) \sim \prod_{i=1}^{d} d_i(t) \sim V(0) \exp\left(\sum_{i=1}^{N} \lambda_i t\right). \tag{2.20}$$

For conservative systems $V(t)$ is constant, and for dissipative systems $\lim_{t\to\infty} V(t) = 0$. Consequently, for chaotic trajectories of conservative systems, the value of the sum of positive exponents has to coincide with the absolute value of the sum of negative exponents; while, for dissipative systems, the value of the sum of all positive Lyapunov exponents has to be smaller then the absolute value of the sum of negative exponents. All this together implies that a typical spectra of Lyapunov exponents of an attractor that is not a fixed point will always have one null exponent, and several negative exponents; if there are positive exponents the attractor is chaotic, if not it is some kind of periodic attractor.

Fig. 2.8 illustrates these ideas for the example of a generic three-dimensional dissipative chaotic system. The Lyapunov spectrum has to be of the type $\{\lambda_1 > 0, \lambda_2 = 0, \lambda_3 < 0\}$ with $\lambda_1 < |\lambda_3|$. An infinitesimal sphere of radius R, centered in the trajectory point $\mathbf{x}(0)$ and starting at $t = 0$, has become an ellipsoid of semi-axes $a > b > c$ a time later, $t > 0$. This is related to the Lyapunov exponents as follows. Along the direction associated to λ_1 (the expanding direction) there is exponential divergence of trajectories; therefore, the semi-axis of the ellipsoid is $a = d_1(t) \sim R{\cdot}\exp(\lambda_1 t) > R$. Along the direction associated to λ_3 (the contracting direction) there is exponential convergence, and the correspondent semi-axis has a length $c = d_3(t) \sim R{\cdot}\exp(\lambda_3 t) < R$. Along the direction tangent to the sphere, associated to λ_2 (the neutral direction) there is not convergence nor divergence of trajectories, then the length of the correspondent semi-axis is $b = d_2(t) \sim R{\cdot}\exp(\lambda_2 t) = R$. Moreover, because $\lambda_1 < |\lambda_3|$ the volume of the ellipsoid has to be smaller than the volume of the sphere: $V(t) \sim a \cdot b \cdot c \sim R^3 \cdot \exp(\lambda_1 - |\lambda_3|) < V(0)$.

Similar examples could had been posed for the other attractors that are possible in three dimensions: the quasi-periodic attractor $\{\lambda_1 = 0,\ \lambda_2 = 0, \lambda_3 < 0\}$, the limit cycle $\{\lambda_1 = 0, \lambda_2 < 0, \lambda_3 < 0\}$, and the fixed point $\{\lambda_1 < 0, \lambda_2 < 0, \lambda_3 < 0\}$. In more than three dimensions there is the possibility of having dissipative hyper-chaotic attractors, these are chaotic attractors that have more than one positive Lyapunov exponent; for a system in a four-dimensional phase space the spectrum of Lyapunov exponents of the hyperchaotic attractor is of the type $\{\lambda_1 > 0, \lambda_2 > 0, \lambda_3 = 0, \lambda_4 < 0\}$ with $\lambda_1 + \lambda_2 < |\lambda_4|$. Besides the hyper-chaotic attractor, in this case, there are two possible types of chaotic attractors $\{\lambda_1 > 0, \lambda_2 = 0, \lambda_3 = 0, \lambda_4 < 0\}$ with $\lambda_1 < |\lambda_4|$, and $\{\lambda_1 > 0, \lambda_2 = 0, \lambda_3 < 0, \lambda_4 < 0\}$ with $\lambda_1 < |\lambda_3| + |\lambda_4|$.

For the calculation of Lyapunov exponents it is considered that if $\mathbf{x}(t)$, which started at $\mathbf{x}(0)$ is given by Eq. 2.1, the vector position of a close orbit, $\mathbf{x}(t) + \delta\mathbf{x}(t)$, started at $\mathbf{x}(0) + \delta\mathbf{x}(0)$, with respect to the $\mathbf{x}(t)$ will verify the so called linearized equations:

$$\frac{d\left[\delta\mathbf{x}(t)\right]}{dt} \approx \left(\frac{\partial\mathbf{F}}{\partial\mathbf{x}}\right)_{\mathbf{x}(t)} \cdot \delta\mathbf{x}(t), \tag{2.21}$$

with $(\partial\mathbf{F}/\partial\mathbf{x})_{\mathbf{x}(t)}$ the Jacobian matrix of the vector field computed along the reference trajectory $\mathbf{x}(t)$, whose components are $[\partial\mathbf{F}/\partial\mathbf{x}]_{\alpha,\beta} = \partial F_\alpha/\partial x_\beta$, with $\alpha, \beta = 1, 2, ..., d$. To obtain the Lyapunov spectra [Benettin et al. (1980b); Wolf et al. (1985)], given an initial condition in the attractor, $\mathbf{x}(0)$, a set of d vectors $\{\delta\mathbf{x}_1(0), \delta\mathbf{x}_2(0), ..., \delta\mathbf{x}_d(0)\}$, each of length equal to one, and all perpendicular to each other is chosen to define initial conditions of close trajectories. For a short time interval, the time evolution, $\mathbf{x}(t)$, of the reference trajectory is computed from $\mathbf{x}(0)$ by means of Eq. 2.1. Simultaneously, for this $\mathbf{x}(t)$, the linearized equation (Eq. 2.21) is integrated d times, each for each $\delta\mathbf{x}_i(0)$, $i = 1, 2, ..., d$, to obtain the time evolution, $\{\delta\mathbf{x}_1(t), \delta\mathbf{x}_2(t), ..., \delta\mathbf{x}_d(t)\}$, of the orthonormal set of vectors, $\{\delta\mathbf{x}_1(0), \delta\mathbf{x}_2(0), ..., \delta\mathbf{x}_d(0)\}$.

In a numerical evaluation of Lyapunov spectra these time evolutions can only be followed for a limited time interval because of the limits to numeric precision and ranges available for numerical values that are imposed by the finite arithmetic used by the computers. This prevents following exponential increases and decrease for a large time interval. Moreover, the initially orthonormal set of vectors loses its orthonormality because all vectors tend to follow the direction of maximum growth. These problems are solved by following the evolution of $\{\delta\mathbf{x}_1(t), \delta\mathbf{x}_2(t), ..., \delta\mathbf{x}_d(t)\}$ for a finite time interval, T, short enough to avoid finite arithmetic problems, and long enough to allow a significant evolution of $\{\delta\mathbf{x}_1(t), \delta\mathbf{x}_2(t), ..., \delta\mathbf{x}_d(t)\}$. Then, the set of vectors, $\{\delta\mathbf{x}_1(T), \delta\mathbf{x}_2(T), ..., \delta\mathbf{x}_d(T)\}$, achieved at the end of this interval is reconverted to an orthonormal set by means of the Gramm–Schmidt orthonormalization method [Riley et al. (1998)], which is a standard technique of linear algebra to obtain a set of orthonormal vectors from a set of non-orthonormal vectors that span the same space. In this method, the vectors at the end of the interval are first made orthogonal to each other to obtain the set of orthogonal vectors $\{\delta\mathbf{X}_1, \delta\mathbf{X}_2, ..., \delta\mathbf{X}_d\}$, which according to Eq. 2.19 provide an estimate of each of the $i = 1, 2, ..., d$ Lyapunov exponents by means of $\lambda_i \approx \ln(\|\delta\mathbf{X}_i\|)/T$, with $\|\cdot\|$ indicating the euclidean norm. Then the set is normalized by dividing each vector by its euclidean norm to obtain the orthonormal vector set $\{\delta\mathbf{X}_1/\|\delta\mathbf{X}_1\|, \delta\mathbf{X}_2/\|\delta\mathbf{X}_2\|, ..., \delta\mathbf{X}_d/\|\delta\mathbf{X}_d\|\}$ ready to repeat the process.

The iterated use of the Gramm–Schmidt method, combined with the tendency of the vectors to align along the direction of maximum growth, has as a consequence that the initial set of vectors $\{\delta\mathbf{x}_1(0), \delta\mathbf{x}_2(0), ..., \delta\mathbf{x}_d(0)\}$, that was an orthonormal set arbitrarily oriented, evolves after successive steps towards an orthonormalized set oriented along the principal axes [Benettin et al. (1980b); Wolf et al. (1985)]. This finally allows the Lyapunov exponents to be computed from a large enough number of iterates as a time average

$$\lambda_i = \frac{1}{N \cdot T} \sum_{k=1}^{N} \ln\left[\left\|\delta\mathbf{X}_i^{(k)}\right\|\right], i = 1, 2, ..., d, \tag{2.22}$$

with N the number of iterates, and $\delta\mathbf{X}_i^{(k)}$ the orthogonalized vectors obtained at the end of the kth step.

An example of study of a Lyapunov spectrum will now be presented for the Rössler model [Rössler (1976)], which is an autonomous chaotic system given by the set of equations

$$dx/dt = -(y + z), \tag{2.23}$$

$$dy/dt = x + ay, \tag{2.24}$$

$$dz/dt = b + z(x - c). \tag{2.25}$$

This was introduced as an example of a very simple flow able to display chaos. It is mathematically simple in the sense that the field $\mathbf{F}(\mathbf{x})$ has only seven terms, being only one of them nonlinear with the nonlinearity, $z \cdot x$, only quadratic. For the parameter values $a = 0.2$, $b = 0.2$, and $c = 5.7$, it has been shown [Rössler (1976)] to display a simple chaotic attractor in which the trajectory rotates around a fixed point. Although, even simpler mathematical attractors were later found by a systematic computational search [Sprott (1994)], the Rössler model has become a favorite system for researchers to perform demonstrations of theoretical results, or to substantiate experimental observations.

To compute the spectra of Lyapunov exponents, Eq. 2.23–2.25 were integrated to obtain the reference trajectory $\mathbf{x}(t) = [x(t), y(t), z(t)]$, and Eq. 2.21 was integrated, using the Jacobian matrix

$$\left(\frac{\partial \mathbf{F}}{\partial \mathbf{x}}\right)_{\mathbf{x}(t)} = \begin{bmatrix} 0 & -1 & -1 \\ 1 & a & 0 \\ z(t) & 0 & x(t) - c \end{bmatrix} \tag{2.26}$$

to obtain the evolution of a orthonormal set of vectors along that trajectory. The values of the Lyapunov exponents change with the parameter values used in the calculation. In this example, $b = 0.2$, and $c = 5.7$, were fixed, and the spectrum was computed as a function of $a \in (0, 3.8)$. The results are displayed in Fig. 2.9.

The three Lyapunov exponents are presented in Fig. 2.9(a): there is a minimum exponent, always negative, which takes values within the range $-6 < \lambda_3 < -3.5$, a null exponent, and an exponent which changes sign as a changes, whose values range between -0.8 and 0.2. An enhanced plot of these last two exponents appears in Fig. 2.9(b). In the range of values of a studied it is observed that: (i) the system is strongly dissipative, because of the relatively large magnitude of the absolute value of the third exponent, (ii) for $a \lesssim 0.0065$ the attractor is a fixed point being the two largest exponents negative: $\lambda_1 \approx \lambda_2 \sim -10^{-2}$, (iii) for $a \gtrsim 0.0065$ the attractor is not a fixed point, because there is a null exponent for all values of a, and (iv) for values of $a \lesssim 0.15$, the attractor is not chaotic (no positive exponents), while for $a \gtrsim 0.15$ the attractor is chaotic (one positive exponent) except in some windows of periodicity. In fact, as a increases from zero, there is a transition to chaotic behavior by means of a period doubling cascade. This is illustrated in Figs. 2.9(c–e) by means of a series of plots of the projection, onto the plane $x - y$, of the attractor followed by the Rössler model for increasing values of the system parameter, a. Fig. 2.9(c) displays

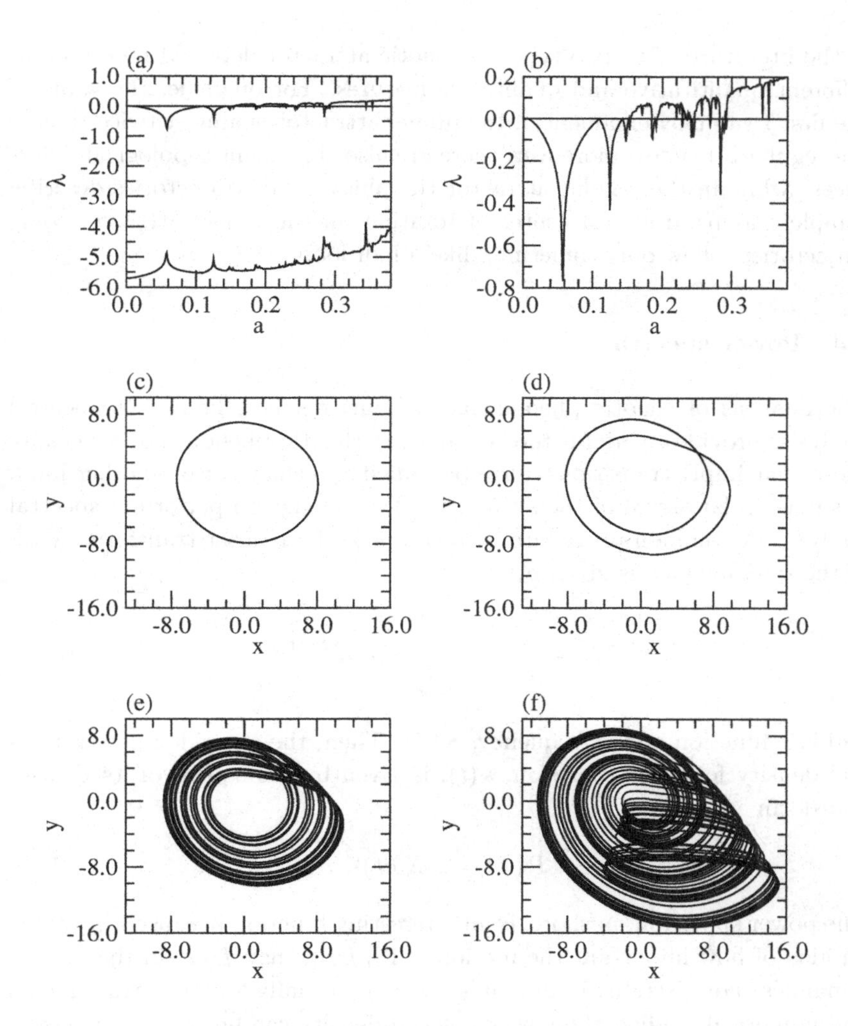

Fig. 2.9 The Lyapunov spectrum of the Rössler attractor as a function of the system parameter a: (a) plot of the whole spectrum, and (b) detailed view of the two largest Lyapunov exponents. Plots of the stable attractor for the system parameter values $b = 0.2$, $c = 5.7$, and (c) $a = 0.07$, (d) $a = 0.14$, (e) $a = 0.18$, and (f) $a = 0.34$.

a limit cycle that occurs for $a = 0.07$, Fig. 2.9(d) shows that for $a = 0.14$ the period of the limit cycle has doubled, and Fig. 2.9(e) shows the chaotic attractor achieved for $a = 0.18$. In Fig. 2.9(f) it is presented the chaotic attractor that occurs at large values of a, in particular $a = 0.34$; this is an example of a form of the Rössler attractor which is called a funnel attractor in the literature. The two types of chaotic attractor depicted here present different quantitative and qualitative features. For one side, the value of the first Lyapunov exponent of the funnel attractor is almost twice that of the regular attractor; moreover, there are also significant topological differences: while in the regular attractor the phase space trajectories describe complete loops around a center of rotation, in the funnel attractor some trajectories follow only something like a half loop.

2.4 Power spectra

The analysis of chaotic phenomena in Fourier space, known as spectral analysis, provides a useful tool to analyze chaotic phenomena, especially those that imply transitions, form periodicity to chaos, or of another kind. A single scalar signal of the system, $s(t)$, is enough to perform a spectral analysis. A common procedure is to compute the Fourier transform, $\mathfrak{F}[x]$, of the signal, which is given by

$$\mathfrak{F}[s] = S(f) = \int_{-\infty}^{\infty} s(t)\, e^{i2\pi f t} dt, \tag{2.27}$$

and is a function of the frequency, $S(f)$. Then, the one-sided power spectral density for a real function, $s(t)$, is given by the square of its Fourier transform

$$P(f) = 2\,|X(f)|^2 . \tag{2.28}$$

The power spectrum, in short, is an interesting function because it provides an idea of how important the motions with frequency f are in the system dynamics. For discrete finite signals which are usually found in experiments and numerical studies, the power spectral density can be straightforwardly computed by means of a standard technique known as the Fast Fourier Transform. This is described in many standard textbooks on numerical analysis [Press et al. (1992)], and is frequently implemented in numerical and graphical software packages.

Two structural features of the power spectrum, sharp peaks and broadband noise, are useful to distinguish chaotic from non-chaotic signals [Eckmann and Ruelle (1985)], as well as to characterize these signals to some extent. The power spectra, $P(f)$, of periodic attractors are made of Dirac δ-peaks at the dominant frequencies of the attractor, f_1, and at its harmonics, $k \cdot f_1$, with k a whole number,

$$P(f) = \sum_{k=1}^{\infty} A_k \delta(k \cdot f_1 - f), \tag{2.29}$$

with A_k a decreasing amplitude, and $\delta(x)$ the Dirac delta function [$\delta(x) \to \infty$ for $x = 0$, and $\delta(x) = 0$ for $x \neq 0$]. This feature of periodic attractors, however, is lost to some extent when the power spectrum is computed from time series that are finite and discrete; but even in this case the power spectrum is significant, because the δ-peaks become acute peaks with finite height and a finite width (of the order of the inverse of the length of the time series), which are usually clearly distinguishable. The main feature of the power spectrum of chaotic attractors is its continuity: it presents a broad noise-like band structure in general, with finite and broad peaks in many systems. These qualitative differences make the power spectrum useful to distinguish between chaotic and non-chaotic signals, to determine the main frequencies of periodic oscillators, and to assign characteristic frequencies to chaotic oscillators. Moreover, the structure of the broad band noise may also be a characteristic feature of a chaotic attractor.

The use of power spectra for analysis of chaotic systems will be illustrated now by means of the Hindmarsh–Rose neuron model [Hindmarsh and Rose (1984)]. This is a phenomenological model designed to reproduce the main features of the electrical activity of a single isolated neuron. The relevant variable is the membrane voltage in the axon, which is the property that characterizes the nerve impulse that transports information along the neuron axon. The behavior of the membrane voltage is determined by the transport of ions across protein channels stuck in the membrane, which can be one of two types: fast or slow. Moreover, there are external outputs, which are given by electric currents injected in the neuron from the environment. The Hindmarsh–Rose equations [Hindmarsh and Rose (1984)] define a model for the dynamics of the membrane voltage, $x(t)$, which in appropriate ranges of parameters has been found to be a realistic description of the electrophysics observed in experiments made with single neurons [Rabinovtich et al. (1997)]. The equations of the model written in dimensionless form are:

$$dx/dt = y + 3x^2 - x^3 - z + I, \tag{2.30}$$

$$dy/dt = 1 - 5x^2 - y, \tag{2.31}$$

$$dz/dt = -r\left[z - 4(x + 8/5)\right]. \tag{2.32}$$

In these equations $y(t)$ and $z(t)$ are auxiliary variables describing, respectively, fast and slow transport processes across the membrane. The external current applied, I, and the internal state of the neuron, r, are the control parameters of the model used often.

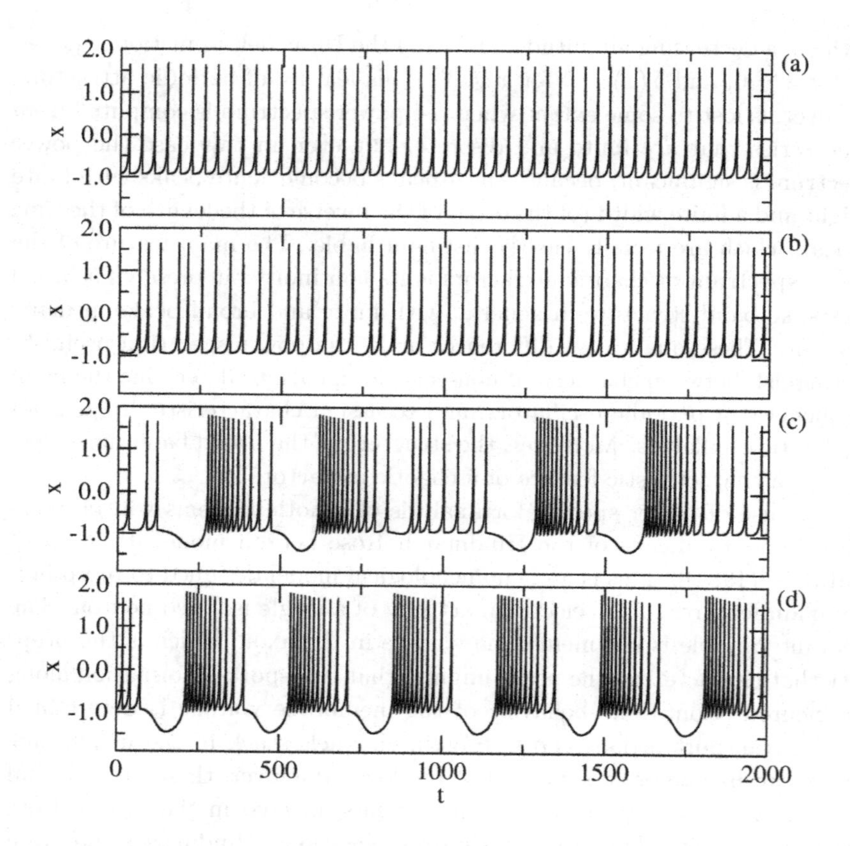

Fig. 2.10 Time series for the membrane voltage, $x(t)$, for the Hindmarsh–Rose model for $r = 0.0021$, and values of I selected to display the relevant dynamic behaviors allowed to this system: (a) periodic spiking ($I = 3.38$), (b) chaotic spiking ($I = 3.31$), (c) chaotic bursting ($I = 3.28$), and (d) periodic bursting ($I = 3.21$).

The Hindmarsh–Rose model is able to display different dynamical regimes observed in experiment: rest, bursts, spikes, with the last two allowing periodic and chaotic dynamics. The rest regime is a very simple one in which the membrane voltage is constant or almost constant (a fixed point, or a low-frequency and low-amplitude limit cycle). The spiking and bursting regimes are displayed in Fig. 2.10. The spiking behavior, as shown in Fig. 2.10(a, b) is characterized by an oscillation of the membrane voltage, which can be viewed as a series of acute peaks, called spikes. This series can be periodic [Fig. 2.10(a)], when the spikes fire regularly at definite time intervals, or aperiodic [Fig. 2.10(b)] when the intervals between two consecutive spikes changes chaotically. The bursting behavior [Figs. 2.10(c, d)] is characterized by the alternation between two types of dynamics: one in which the systems appears to be at rest, disrupted by another made of bursts, which are time intervals of spiking behavior. Chaotic [Fig. 2.10(c)] and periodic [Fig. 2.10(d)] versions are also possible. In the periodic regime all bursts have the same number of spikes and appear at a periodic pace, in the chaotic version the bursts are irregular regarding both their appearance and the number and rate at which the spikes are fired. The transition between these two relevant dynamical behaviors when a control parameter is changed has been found to be quite sudden [González-Miranda (2003)].

The power spectra that correspond to the time series in Fig. 2.10 appear in Fig. 2.11. For the spiking periodic signal the characteristic structure of sharp peaks at the fundamental frequency ($f_H \approx 0.026$) and its harmonics (two of them appear in the figure) is displayed in Fig. 2.11(a); while for a chaotic spiking regime, close to the above periodic state, a similar structure of peaks (slightly sifted to low frequencies) emerges from a broadband noise-like structure [Fig. 2.11(b)]. The frequency f_H is the frequency with which the spikes are fired in the periodic regime, and is practically equal to the average frequency of firing in the chaotic regime. For the chaotic bursting regime the power spectrum has lost all peaks, and provides an example of a continuous almost flat structure [Fig. 2.11(c)] proper of chaotic systems; however, a broad maximum has emerged at low frequencies ($f_L \approx 0.003$) and some bumps, reminiscent of the main peak appear around $f_H \approx 0.026$. Finally, Fig. 2.11(d) shows the power spectrum for the periodic bursting regime, which when compared with Fig. 2.11(a), shows that a new series of peaks made of a fundamental low frequency ($f_L \approx 0.003$) and its harmonics has appeared. This happens to be compatible with the simultaneous existence of the above high frequency series of peaks ($f_H \approx 0.026$). In the bursting regime there is a dynamics that occurs at two time scales that differ in an order of magnitude: the scale at which the bursts occur with a characteristic time $T_L = 1/f_L \approx 333$, and the scale of the spikes whose characteristic time is $T_H = 1/f_H \approx 38.4$.

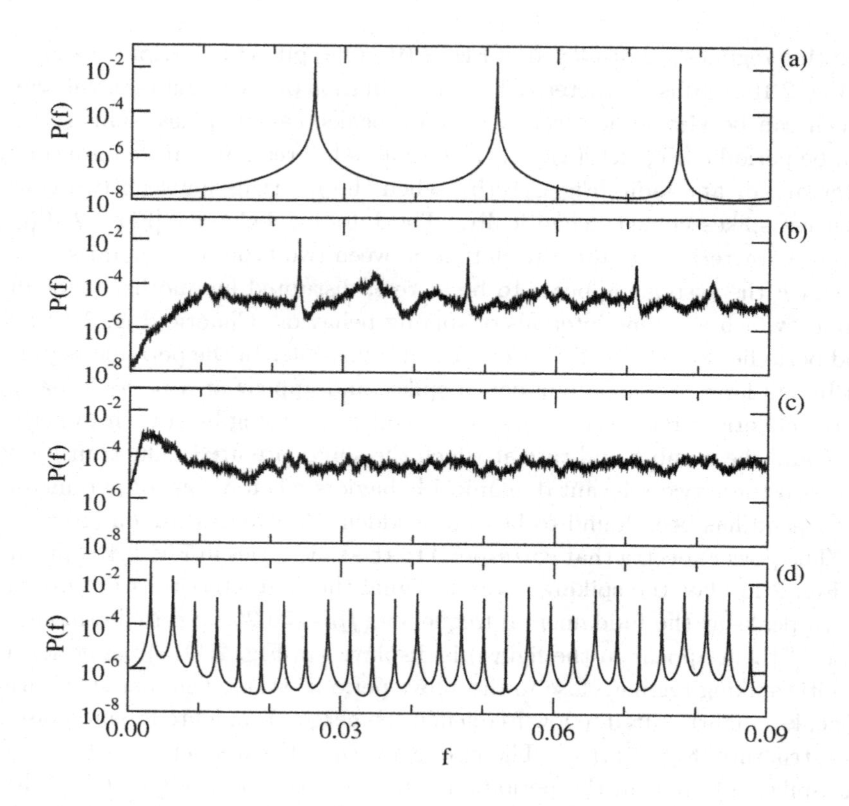

Fig. 2.11 The power spectral density obtained from $x(t)$ for the Hindmarsh–Rose model for the system parameter values $r = 0.0021$, and (a) $I = 3.38$, (b) $I = 3.31$, (c) $I = 3.28$, and (d) $I = 3.21$.

2.5 Unstable periodic orbits

Theoretical reasoning [Cvitanovic 1988; Grebogi et al. (1988)], substantiated by numerical evidence [Auerbach et al. (1987); Pawelzik and Schuster (1991)], and laboratory experiments [Badii et al. (1994)] sustain the assertion that a chaotic attractor has embedded an infinity of unstable periodic orbits. A particular chaotic trajectory, then results from an aperiodic succession of motions in the neighborhoods of a number of these unstable periodic orbits. The shapes and distribution of unstable periodic orbits determine the structure of the chaotic attractor; in this sense it is said that they constitute the attractors skeleton. From the knowledge of the properties of the periodic orbits of a given attractor it is possible to calculate invariant properties, such as Lyapunov exponents, as well as to analyze many interesting chaotic phenomena.

Numerical techniques have been developed to extract a number of periodic orbits from the time series of the dynamical variables which describe the state of the system, $\mathbf{x}(t)$, obtained from the equations of motion or from experiments. These are extracted hierarchically: first, period one orbits, then period two orbits, and so forth up to orbits of order p. Usually a chaotic attractor can be well approximated by the set of unstable orbits so obtained.

A basic and intuitive algorithm for discrete maps [Auerbach et al. (1987)] proceeds as follows. It is assumed that a series of points of the map, $\{\mathbf{x}_1, \mathbf{x}_2, ..., \mathbf{x}_N\}$, large enough as to all the periodic orbits up to the largest order of interest are visited, is known. To extract the orbits with a periodicity of order p, the algorithm proceeds in three steps.

The first step is to extract all the motions that occur in the neighborhood of unstable periodic orbits of order p. To do this a distance r several orders of magnitude smaller than the largest distance between points of the attractor, $\max\{|\mathbf{x}_i - \mathbf{x}_j|\}$, is chosen. Then, all possible sequences of p consecutive data points, $S_p = (\mathbf{x}_i, \mathbf{x}_{i+1}, ..., \mathbf{x}_{i+p})$, are examined and all those which verify $|\mathbf{x}_{i+p} - \mathbf{x}_i| < r$ are considered to be portions of a trajectory in the proximity of an orbit of period p.

The second step is to check the possibility that there may be different types of orbits of period p; and, if this were the case to group the above sequences according to the particular type of orbit to which they belong. To do this a second distance, $R \gtrsim r$, is chosen. Then, a given sequence S_p, is paired with each of all the other sequences, S'_p, and, for each pair, the distances $D_P = |S_p - P(S'_p)|$, with $P(\cdot)$ changing over the p possible cyclic permutations of the elements of S'_p, are computed. If the minimum D_P verifies $D_P < R$, the two sequences are considered to belong to the same unstable periodic orbit. When all the sequences that belong to the same orbit that S_p have been found they are brought apart, and the process is repeated with the remaining sequences. Finally all sequences are grouped in classes, such as $C_p = \{S_p, P(S'_p), ...\}$, with $P(S'_p)$ the permutation that gave the minimum D_P; each class contains all the approximations available for the correspondent unstable periodic orbit.

The third step is to obtain an approximation of the unstable periodic orbit. This is simply made by the calculation of the center of mass of all data points of the sequences contained in each C_p. Finally, each of the periodic orbits so obtained, which has a period greater than two, has to be examined to see if it is made of two or more periodic orbits of smaller periodicity; if this is the case it is discarded. The remaining orbits constitute the set of unstable periodic orbits extracted from the time series.

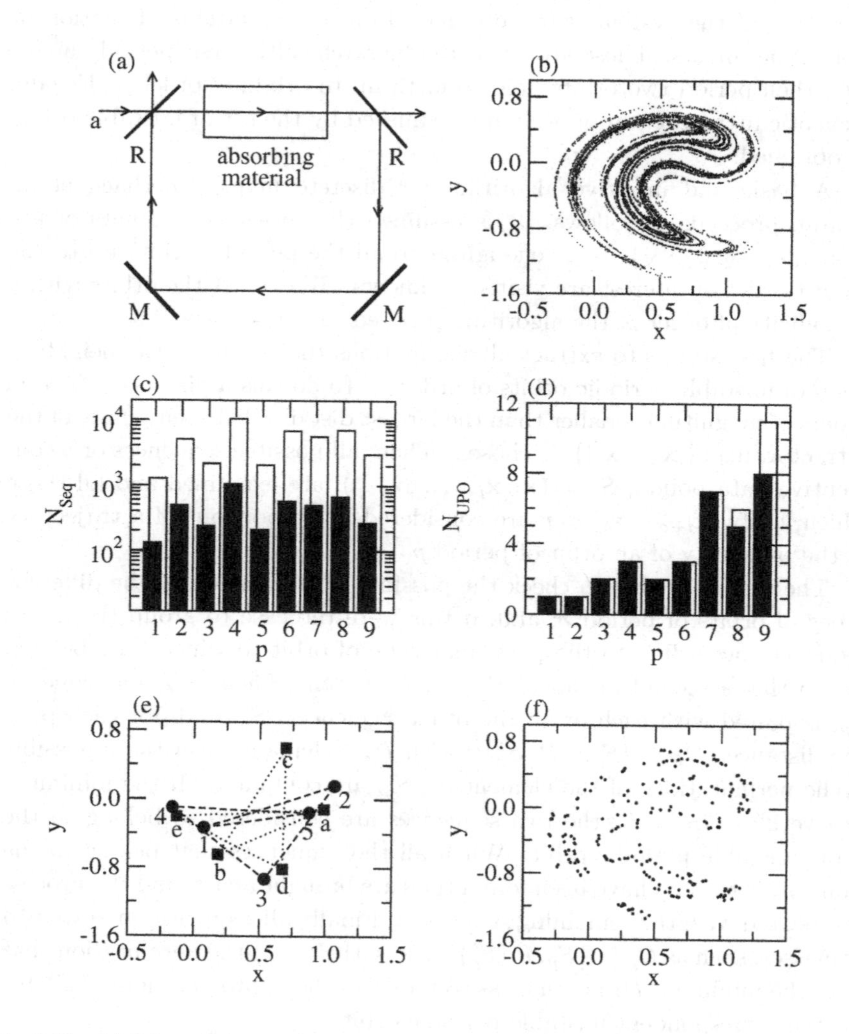

Fig. 2.12 (a) Scheme of an optical ring cavity system. (b) The Ikeda chaotic attractor. (c) Number of periodic sequences of length p, and (d) number of unstable periodic orbits of period p obtained from time series of 10^6 (filled bars) and 10^7 (hollow bars) points. (e) The two orbits of period 5: the arrow head (circles and dashed lines), and the star (squares and dotted lines). (f) The 36 unstable periodic orbits obtained up to period 9.

These ideas will be illustrated by means of its application to a two-dimensional discrete map, known as the Ikeda map in the literature [Ikeda (1979); Hammel et al. (1985)]. The equations of the map are

$$x_{j+1} = \alpha + R \cdot (x_j \cos\theta_j - y_j \sin\theta_j)\,, \tag{2.33}$$

$$y_{j+1} = R \cdot (x_j \sin\theta_j + y_j \cos\theta_j)\,, \tag{2.34}$$

with

$$\theta_j = \phi - \frac{p}{1 + (x_j^2 + y_j^2)}\,, \tag{2.35}$$

and a, R, p, and ϕ the system parameters. These equations model the dynamics of the electric field in a device of interest in nonlinear optics known as a ring cavity which is made by a set of mirrors arranged in such a way that a beam of light can be made to follow a closed path. This is done [Fig. 2.12(a)], for example, by means of four mirrors located in the corners of a rectangle, and placed in such a way that a light beam falling upon one of them with an angle of 45° will follow a trip along the sides of the rectangle to return to the first mirror with the same angle after completing a cycle. Two of the mirrors are completely reflective (M), while the other two are beam splitters (R), which transmit a fraction of the incident light, so that they can be used to feed the system, and to obtain a measurable output. Between these two mirrors there is a sample of a nonlinear optical absorber material [Ikeda (1979)] which interacts with the light beam. The dynamical variables x and y are the dimensionless real and imaginary part of the amplitude of the complex electric field in the cavity. The time is discrete because this field is modified once per cycle; i.e. x_j and y_j are supposed to be measured at the cycle number j of the light in the ring. The parameter α measures the input amplitude field, R is the reflectivity of the mirrors, and ϕ and p characteristic properties of the cavity and the absorber [Hammel et al. (1985)].

For the particular example presented here the values of the parameters in Eqs. 2.33–2.35 are $a = 0.80$, $R = 0.90$, $p = 0.54$, and $\phi = 0.60$ which lead to motion in a chaotic attractor [Hammel et al. (1985)]. This is displayed in Fig. 2.12(b). The above method to extract periodic orbits from the time series has been applied to two large time series of 10^6 and of 10^7consecutive phase space points. The number of periodic sequences of lengths, p, ranging from 1 to 9, obtained after performing the first step of the extraction process on the time series of 10^6 points had orders of magnitude between 10^2 and of 10^3 depending on the particular length considered. When the time series of 10^7 points was studied these figures increased by an order of magnitude [Fig. 2.12(c)]. After performing the second and third steps a relatively small number of unstable periodic orbits was obtained [Fig. 2.12(d)]. For periodicities between 1 and 7, the result was independent of the length of the time series which suggests that all unstable periodic orbits have been obtained. The increase of the size of the time series by an order of magnitude allowed the discovery of one new orbit of periodicity $p = 8$ and three new orbits of periodicity $p = 9$. To present an example of a set of unstable periodic orbits, the two orbits of period $p = 5$ are presented in Fig. 2.12(e): one of them, indicated as the sequence $(1, 2, 3, 4, 5)$, has the shape of an arrow head pointing down, while the other, (a, b, c, d, e) , looks like an irregular five point star. Finally, the points pertaining to all 36 unstable periodic orbits obtained are shown together in Fig. 2.12(f); they provide a sketch of the skeleton of unstable periodic orbits embedded in the chaotic attractor shown in Fig. 2.12(b).

The algorithm presented above to obtain unstable periodic orbits assumes a discrete time evolution where periodicity of order p means the repetition of the system state after p steps. For time-continuous systems the extraction of periodic orbits is more difficult, because even for a period one orbit there are many points until a given phase space point is repeated. Even so, conceptually simple algorithms, which generalize the above results, have been developed to extract the dominant unstable periodic orbits from time-continuous systems [Pawelzik and Schuster (1991); Franceschini et al. (1993)]. In one of them [Pawelzik and Schuster (1991)] a technique of counting returns from segments of the orbits allows to determine the length of the periodic orbits; then for each length, the three steps described above are applied to obtain the unstable periodic orbits. The other method resorts to the construction of Poincaré maps [Franceschini et al. (1993)] to convert the time-continuous systems to a discrete map in which the above technique is applied to detect the periodic orbits, then a Newton–Raphson iteration scheme [Press et al. (1992)] is used to reconstruct the whole time-continuous orbit.

More abstract and elaborate techniques have been developed recently for efficient and systematic detection of unstable periodic orbits in chaotic dynamical systems [Schmelcher and Diakonos (1997); Pingel et al. (2001)]. The central idea is to convert the unstable periodic orbits of a chaotic system to stable periodic orbits of a linearly related system. The numerical calculation of the stable periodic orbits is done by integration of the equations of the new system. The method was initially implemented for two-dimensional maps [Schmelcher and Diakonos (1997)], and later generalized to be used with flows [Pingel et al. (2001)].

2.6 Time series analysis

In many experiments of interest in the study of chaotic oscillators, the values of a continuous dynamical variable, $s(t)$, are sampled with a given time interval, Δt. The output of these experiments is, then, a scalar time series of values of s: $\{s(t_1), s(t_2), ..., s(t_N)\}$ given at times $t \in \{t_1, t_2, ..., t_N\}$. The dynamics of chaotic flows occurs in phase spaces with dimensions, d, greater than two. Being the time series the only available information, the problem of the possibility of a reconstruction of the dynamics of the system in a finite-dimensional phase space can be posed [Packard et al. (1980)]. The aim of such reconstruction would be to use the time series to construct a succession of states in a reconstructed space of dimension $d_E \geq d$, which follows a dynamics equivalent to that of the original system in its phase space of dimension d. Here equivalent means that the reconstructed dynamics can be used for the same purposes as that of the original, such as the determination of the topology of the attractor, or the calculation of the Lyapunov exponents. This reconstructed space is known as the embedding space, and its dimension is the embedding dimension.

The possibility of such reconstruction is based [Abarbanel et al. (1993); Schreiber (1999)] on the idea that nonlinear oscillators are deterministic systems. This implies that the present dynamical state of the system is unambiguously determined for the present value of the measured variable, together with a large enough number of values of the same time series in the past. So, given $\{s(t_1), s(t_2), ..., s(t_N)\}$, an approximation to this state is then provided by a vector with components

$$\mathbf{s}(t) = \{\, s(t),\ s(t-\tau), ..., s[t-(d_E-2)\cdot\tau], s[t-(d_E-1)\cdot\tau]\} \quad (2.36)$$

where τ is a time interval known as the time lag. If the embedding dimension, d_E, and the time lag, τ, are properly chosen a succession of states like this follows a dynamics equivalent to the original.

Takens embedding theorem [Takens (1981); Sauer et al. (1991)] states the sufficient condition for the dimension of the embedding space to be appropriate; this is that the relation $d_E \geq 2\ d + 1$ has to be fulfilled. This is a sufficient but not a necessary condition; that is, there can be proper reconstructed dynamics within a space of dimension d_E, such that

$2\,d+1 > d_E \geq d$. When the dimension of the dynamics to be reconstructed is known, Takens theorem provides a criterion that warrants the proper choice of embedding dimension. When the original dynamics are unknown this theorem warrants that an embedding dimension exists.

The choice of the time lag, τ, which allows a proper reconstruction, has to be made to verify two competing requirements. For one side, the time interval τ, has to be large enough to guarantee that the information on the dynamics contained in each of two consecutive elements in the state vector (Eq. 2.36) is different enough so as to allow to consider them as two different coordinates of the reconstructed space. For the other, τ has to be small enough to avoid the deterministic relation between consecutive elements to be lost because of the instability that is characteristic of a chaotic evolution. The embedding theorem tells nothing about the choice of τ, so additional criteria have been developed to this aim [Casdagli et al. (1991); Abarbanel et al. (1993)]. An example, picked up from the several criteria proposed to this aim, is the calculation of the linear autocorrelation between two elements of the time series, $\{s(t_1), s(t_2), ..., s(t_N)\}$, as a function of the time, T, between two of them. This is computed by means of

$$C_L(T) = \frac{\langle S(t+T) \cdot S(t) \rangle}{\langle S^2(t) \rangle}, \tag{2.37}$$

with $S(t_i) = s(t_i) - \langle s(t) \rangle$, and with the angular brackets denoting an average over time $\langle X(t) \rangle = \left[\sum_{j=1}^{N} X(t_j) \right] / N$. The time lag is then chosen as the first value of T, which cancels $C_L(T)$. This choice for τ provides an estimate of the smallest value of the time lag for which two consecutive coordinates of the reconstructed state vector, $\mathbf{s}(t)$, are linearly independent. This is, at least, a proper choice for a first estimate of τ.

Once d_E and τ have been chosen, a sequence of reconstructed states $\mathbf{s}(t)$ can be composed and used to analyze the dynamics of the system that generated the time series, $s(t)$. This can be done using the same algorithms and techniques that would had been used if the real dynamics, $\mathbf{x}(t)$ were known [Abarbanel et al. (1993); Schreiber (1999)].

These ideas on phase space reconstruction will be illustrated here by means of a model of chemical chaos introduced [Gaspard and Nicolis (1983); Nicolis (1990)] to study fundamental aspects of the emergence of chaotic behavior in chemical reacting systems. In particular, these authors studied a form of chaos, know as Sil'nikov homoclinic chaos, in which the structure of a three-dimensional chaotic attractor is given by the competition between an unstable dynamics along a curve, and a stable dy-

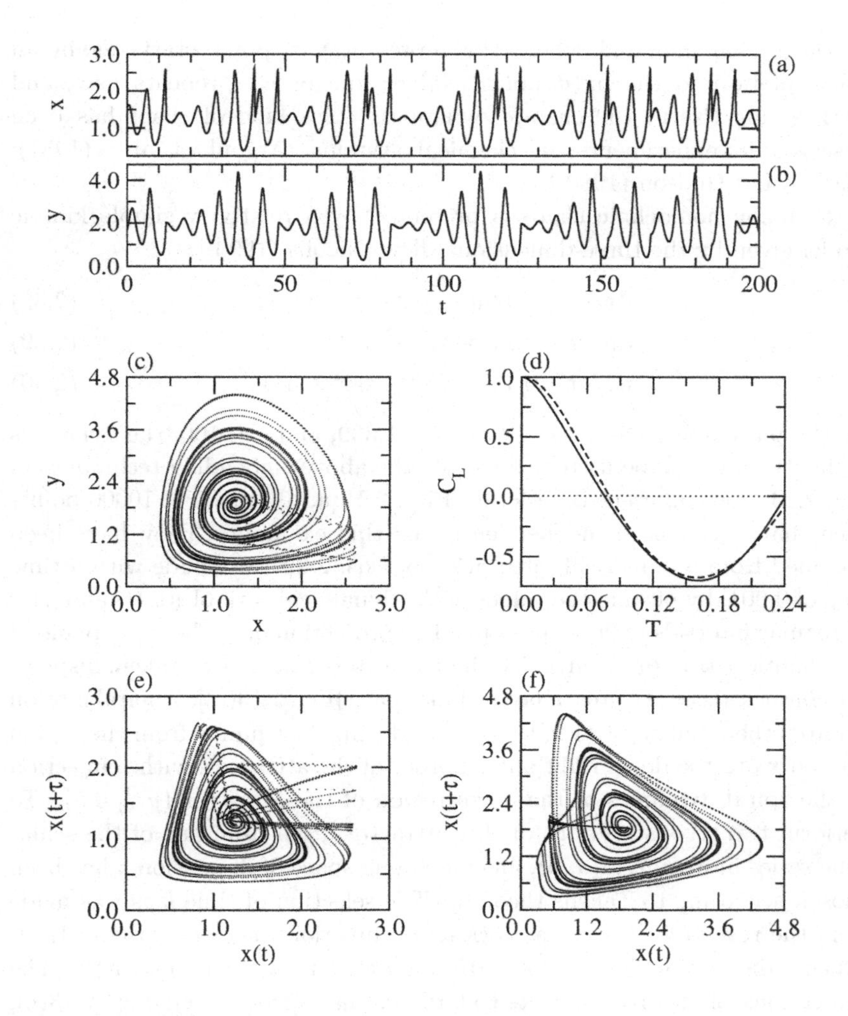

Fig. 2.13 Time series obtained from the time evolution of (a) $x(t)$, and (b) $y(t)$ of the chemical chaos model by Gaspard and Nicolis. (c) Projection of the thee dimensional attractor. (d) Linear self-correlation of $x(t)$ (solid line) and $y(t)$ (dashed line). Projections of the reconstructed dynamics made from (e) $x(t)$, and (f) $y(t)$.

namics on a surface. These two manifolds are tangent in certain fixed point. In the two-dimensional manifold, and far from the fixed point, a representative phase space point spirals down towards the fixed point; when in the vicinity of the fixed point, the effect of the unstable manifold is to throw the point away of the fixed point to a region close to the stable manifold where the representative point starts again an spiral motion to the fixed point. The same process repeats back and forth giving rise to the chaotic dynamics. This kind of chaos has been observed in experiments on chemical systems [Argoul et al. (1987); Bassett and Hudson (1988)].

Sil'nikov homoclinic chaos is presented by a relatively simple kinetic model given by the three-dimensional flow [Nicolis (1990)]

$$dx/dt = x\,(\alpha y - y/2 - z + 3/5)\,, \tag{2.38}$$

$$dy/dt = y\,(x + 3z/10 - \beta)\,, \tag{2.39}$$

$$dz/dt = r\,\left(x - z^3/2 + 3z^2 - 5z\right). \tag{2.40}$$

for the parameter values: $\alpha = 0.51$, $\beta = 1.339$, and $r = 100$, the dynamics is chaotic, and will be used to illustrate the above embedding techniques in Fig. 2.13. Two different time series [Figs. 2.13(a, b)], made of 10000 points each, have been used independently for this example. They have been obtained from a numerical integration of Eqs. 2.38–2.40 made with a time step of 0.001, by picking the values of the dynamical variables $x\,(t)$ and $y\,(t)$ at regular intervals of 20 time steps. The projection onto the $x - y$ plane of the chaotic attractor, from which these time series have been taken, displays the characteristic features of homoclinic chaos [Fig. 2.13(c)]: a spiral motion towards the point $\mathbf{x}_0 = (1.2, 1.8, 0.0)$, and some fast jumps from the region around $\mathbf{x}_0$ to the down right outer border of the attractor, with reinjection to the spiral dynamics in the down region of the attractor ($y \lesssim 0.5$). To perform two reconstructions of this attractor, each from one of the scalar time series in Figs. 2.13(a, b), an embedding space of dimension 5 has been chosen according to Takens theorem. The selection of time lags was made from the results for the linear correlation functions, shown in Fig. 2.13(d) which indicate time lags of $\tau \approx 0.076$ for $x\,(t)$, and $\tau \approx 0.080$ for $y\,(t)$. The projections of the reconstructed attractor made from $x\,(t)$, Fig. 2.13(e), and from $y\,(t)$, Fig. 2.13(f), show the same qualitative features that the original attractor: spirals toward a fixed point and jumps to the border of the attractor. Although the reconstructed attractors appear deformed and rotated respect to the original, the essential topological features associated with homoclinic chaos prevail.

Chapter 3

Periodically Driven Chaotic Oscillators

Theoretical and experimental investigations on the dynamics of chaotic systems driven by a weak external periodic force have revealed the existence of two relevant phenomena: the synchronization of the phase of the chaotic oscillator to that of the applied force, and the suppression of chaos for appropriate weak resonant forces. These will be studied in this chapter. First, the concept of the phase of a chaotic system will be introduced, and its dynamics under periodic driving will be studied for low dimensional chaotic flows, showing the phenomenon know as phase synchronization. Then, the phenomenon of chaos suppression in nonlinear one-dimensional periodically forced oscillators driven by a second weak periodic resonant force will be reviewed. Experimental and numerical observations of these two different phenomena in systems such as discharge plasmas, continuous flow stirred chemical reactions, magneto-mechanical systems, lasers, Josephson junctions, and single neuron models will be discussed.

3.1 Phase synchronization

A periodic force is characterized by its maximum strength, A, and its frequency, ω. The phase of the force for any time t is simply ωt. Synchronization of the phase of a periodically driven chaotic system means that, in some way, the chaotic oscillator evolves in pace with the force; i.e., the phase of the oscillator becomes modified to follow the phase of the force. Besides its interest as a fundamental phenomenon it appears interesting in many situations found in science and engineering, where the dynamics of an ensemble made of similar, but not identical chaotic oscillators, would result in an incoherent collective behavior. A weak external applied force may introduce, by means of the phenomenon of phase synchronization, certain coherence in the collective behavior. For example, this appears useful in biology to understand the development of externally imposed rhythms, such as the circadian rhythms [Winfree (1980)]; or in electrical engineering where the integration of multiple elements in a single unit requires the dynamics of the different elements to be paced, this occurs particularly in multichannel chaotic communications [Yalçinkaya and Lai (1997)].

3.1.1 *Phase of a chaotic oscillator*

The concept of the phase of a chaotic oscillator is not a simple one as it is the concept of the phase of a periodic oscillator; therefore, it needs some study before phase synchronization is addressed. Several definitions of the phase and of the frequency of the oscillator have been proposed.

Most of the work done is for chaotic attractors that follow a proper rotation [Yalçinkaya and Lai (1997)], which is a case frequently found in practice. This happens when the trajectory of the oscillator in phase space describes a rotation around a unique center and along a well defined direction, either clockwise or counter-clockwise.

A straightforward definition [Pikovsky et al. (1997a); Pikovsky et al. (2000)] relies on the fact that, if the attractor follows a proper rotation in phase space, it is possible to find a plane in which the projection of a trajectory in the attractor follows a proper rotation too. Then an intuitive definition of the phase, $\phi(t)$, is the polar angle of the trajectory measured in a Cartesian coordinate system (x, y) in the plane given by the coordinates $x(t)$ and $y(t)$ of the trajectory projection through

$$\phi(t) = \arctan\left[\frac{y(t) - y_0}{x(t) - x_0}\right], \tag{3.1}$$

with (x_0, y_0) the center of rotation of the trajectory projection. In this definition ϕ is a continuous real variable defined in the real line, $-\infty < \phi < \infty$, rather than on the circle, $0 \leq \phi < 2\pi$; therefore, phase values separated by 2π are considered different (some authors call this a lift of the phase variable [Rosa et al. (1998)].) The meaning of this definition is clear as it is a direct generalization of the notion of the phase of the motion in a periodic plane orbit. Moreover it is easy to work with it, provided that the projections $x(t)$ and $y(t)$ can be obtained.

The function $\phi(t)$ so defined increases monotonously to infinity for counter-clockwise rotations, or decreases to minus infinity for clockwise rotations. The dynamics usually found is of the type

$$\phi(t) = \phi_0 \pm \Omega t + \xi(t), \tag{3.2}$$

with ϕ_0 a constant given by the initial conditions, Ω a positive constant defining a constant drift of the phase, and $\xi(t)$ a small bounded chaotic fluctuation having a zero mean. The plus sign is for counter-clockwise rotations and the minus sign for clockwise rotations. The constant Ω is a characteristic angular frequency of the attractor. Some authors propose to estimate it by means of the calculation of an instantaneous frequency

$$\Omega(t) = \frac{d\phi(t)}{dt}, \tag{3.3}$$

which for a proper rotation will be either positive or negative for each value of t. This is properly time averaged, $\langle\Omega\rangle = \langle d\phi/dt\rangle$, to characterize the phase behavior of the oscillator by means of a single number which quantifies the frequency of the oscillator $\Omega = |\langle\Omega\rangle|$. Attention has to be paid to the words 'properly time averaged' because, depending on the shape of the distribution of the values of $\Omega(t)$ a simple arithmetic mean might not be the best estimate. Alternatively, Ω, can be obtained from the slope of a least squares fit to a straight line of $\phi(t)$, which provides $\Delta\phi = \phi - \xi = \phi_0 \pm \Omega t$.

These and the following ideas on the concept of the phase of a chaotic attractor will be discussed and illustrated in this subsection by means of the specific example of the Van der Pol–Duffing oscillator [King and Gaito (1992); Gomes and King (1992)] whose equations of motion are:

$$\dot{x} = -\gamma\left(x^3 - \alpha x - y\right), \tag{3.4}$$

$$\dot{y} = x - y - z, \tag{3.5}$$

$$\dot{z} = \beta y. \tag{3.6}$$

Where γ, α and β are the parameters of the model. Our examples are for the parameter values $\gamma = 100$, $\alpha = 0.35$ and $\beta = 650$ which are such that the oscillator dynamics, when started from initial conditions in the appropriate basin of attraction, falls down to a chaotic attractor which follows a proper rotation around the point $(\sqrt{\alpha}, 0, \sqrt{\alpha})$ [King and Gaito (1992); Gomes and King (1992)].

The above ideas on the phase of a chaotic oscillator are illustrated in Fig. 3.1. In Fig. 3.1(a) a trajectory which starts in a point in the attractor is projected onto the $x - y$ plane where it cycles clockwise three times around the center $(x_0, y_0) = (\sqrt{\alpha}, 0)$. The phase angle, ϕ, defined in Fig. 3.1(a),

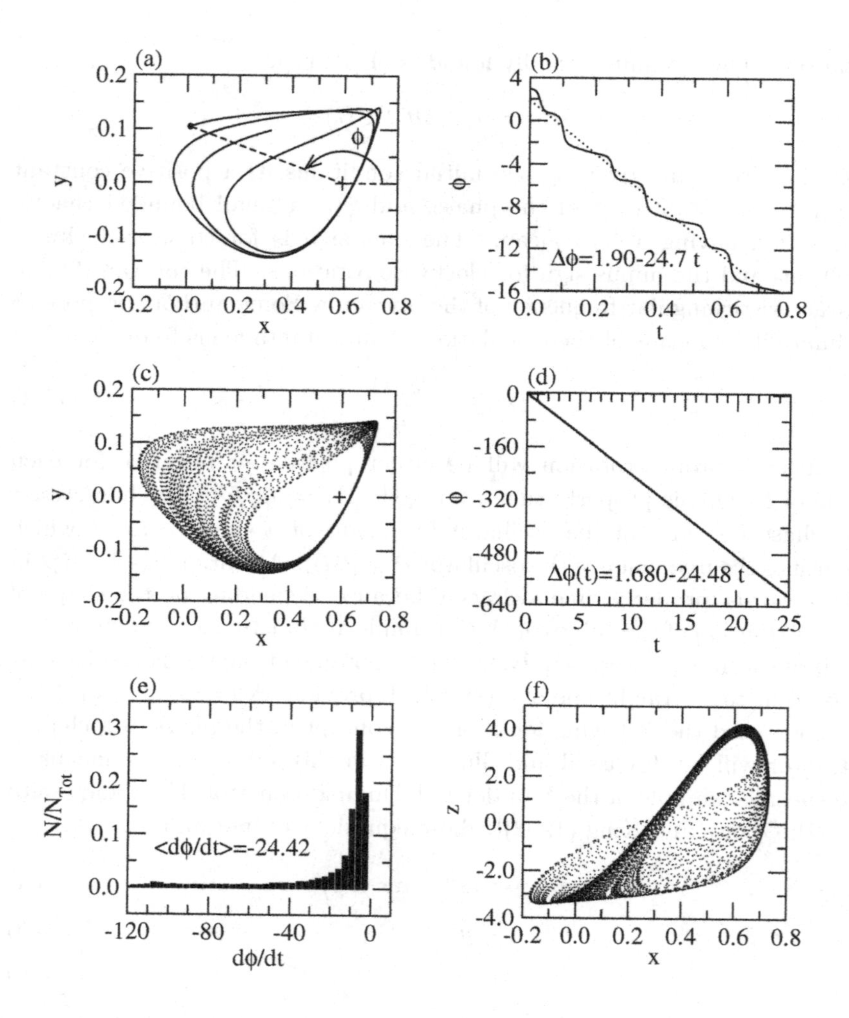

Fig. 3.1 The phase of a chaotic oscillator. (a) Projection onto the plane $x - y$ of a short trajectory, whose center is in the cross, and starting point in the big dot (having a phase ϕ). (b) Time evolution of the phase (solid line), and its fit to a straight line (dashed line). (c, d) The same as (a, b) for a large time interval. (e) Statistics of the derivative of the phase. (f) Projection of the attractor onto the plane $x - z$.

follows a time evolution that can be described by Eq. 3.2, as shown in Fig. 3.1(b). A least squares fit of $\phi(t)$ to a straight line provides, $\Delta\phi$, and then the estimate for the frequency: $\Omega = 24.7$ (see the lower left corner of the figure). Only three loops provide poor statistics, more reliable results are obtained by following the system for 100 loops as done in Figs. 3.1(c) and 3.1(d). The least squares fit estimate for the frequency is now better: $\Omega = 24.48$. The calculation of the frequency as an average of the values of $\Omega(t)$ provided by Eq. 3.3 is illustrated in Fig. 3.1(e) which shows a histogram of the values obtained for $\Omega(t)$ from the time evolution in Fig. 3.1(c), all the values obtained are negative(the tail of the distribution, which extended up to $d\phi/dt = -162.5$, is not fully displayed for the clarity of the figure) and its arithmetic mean gives a value of Ω in an agreement better than 0.25 percent with the obtained from a least squares fit. Finally, having seen a positive, a negative is illustrated in Fig. 3.1(f): the projection of the same attractor onto the $x - z$ plane does not provide a proper rotation because in some attractor regions the rotation is counter-clockwise (top right) while in others it is clockwise (bottom left).

The use of a second practical, although not so intuitive definition of the phase of a chaotic oscillator, borrowed from the field of signal processing [Gabor (1946)], has been proposed [Pikovsky et al. (1997a)]. This can be applied when only a scalar state variable of the system, $s(t)$, is available. The corresponding phase, $\phi(t)$, is then the argument of the analytic signal, which is given by the complex number

$$\sigma(t) = s(t) + i s_H(t), \tag{3.7}$$

with $s_H(t)$ the Hilbert transform of $s(t)$ defined by

$$s_H(t) = \frac{1}{\pi} \lim_{\varepsilon \to 0} \left[\int_{-\infty}^{t-\varepsilon} \frac{s(t')}{t-t'} dt' + \int_{t+\varepsilon}^{\infty} \frac{s(t')}{t-t'} dt' \right], \tag{3.8}$$

which is an improper integral computed according to the principal value of Cauchy. In this case the scalar signal, $s(t)$, is converted to a motion in the complex plane, $[s(t), s_H(t)]$ and the definition of the phase, $\phi_H(t)$, is made in this plane by means of an equation like Eq. 3.1 with the variables x and y substituted by s and s_H. The calculation of the improper integral to obtain, $s_H(t)$, can be done by means of Fourier analysis noting that the integral of Eq. 3.8 is the convolution of the signal, $s(t)$, with the function $f(t) = 1/t$. Therefore, the following expression can be used:

$$s_H(t) = \mathfrak{F}^{-1}\left[i \cdot sgn(f) \cdot S(f)\right], \tag{3.9}$$

with $\mathfrak{F}^{-1}$ the inverse Fourier transform operator, $S(f)$ the Fourier transform (Eq. 2.27) of $s(t)$, and $sgn(f)$ is the function sign of f. Standard techniques of digital filtering can then be used to obtain $s_H(t)$ from $s(t)$

[Jackson (1996)]. Because the final result is a rotation in a the plane $s - s_H$, instead of the plane $x - y$, the same considerations made above regarding the dynamics of the phase, $\phi_H(t)$, and the calculation of a frequency, Ω_H, do apply when using the Hilbert transform.

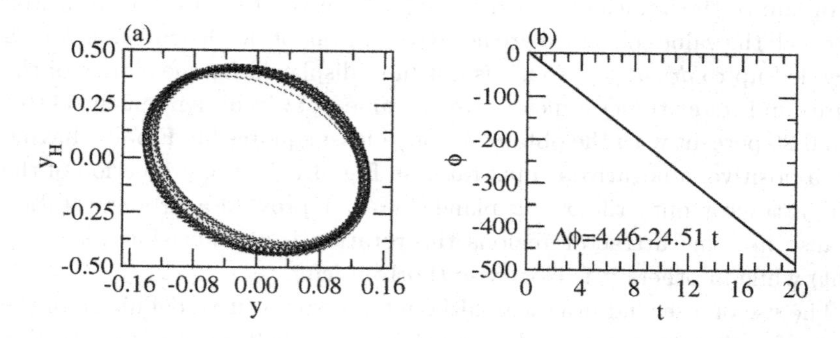

Fig. 3.2 The phase of a chaotic attractor obtained from the analytic signal approach illustrated by the Van der Pol-Duffing oscillator. (a) Evolution of the analytic signal in the complex plane $y - y_H$. (b) Time evolution of the phase.

The illustration of these ideas by means of the example of the Van der Pol–Duffing oscillator is presented in Fig. 3.2. Here the signal used is $\sigma(t) = y(t)$, which fluctuates around zero, and its Hilbert transform was obtained by means of of Eq. 3.9, using the technique of the Fast Fourier transform [Press et al. (1992)]. The analytic signal moves along a clockwise proper rotation around the origin of coordinates [Fig. 3.2(a)] of the $y - y_H$ plane, and the dynamics of its phase [Fig. 3.2(b)] provides a frequency for this system of $\Omega_H = 24.51$ which is very close to others obtained from the dynamics in the $x - y$ plane.

An alternative to be used for a single scalar state variable is to reconstruct the whole attractor by means of standard embedding techniques [Abarbanel et al. (1993); Schreiber (1999)], as discussed in Chapter 2, and proceed with the reconstructed attractor as if the dynamics in the full phase space were known.

A third approach to the study of the phase of a chaotic oscillator uses stroboscopic plots of the attractor [Rosa et al. (1998)]. In this case the values of the system variables at fixed time intervals, τ, are recorded and plotted. If the system were periodic, with period T, the plot of these points will, in general, depict the system trajectory with a non-numerable set of points, when τ/T is an irrational number, a numerable set of points, when τ/T is rational, and a single point when τ/T is a whole number (in particular, this occurs when $\tau = T$). For the chaotic attractor, assuming a proper rotation in a given plane, the stroboscopic plot for arbitrary τ will be a scatter of plots restricted to the intersection of the phase space region spanned by the attractor with the plane where it is projected. But for $\tau = 2\pi/\Omega$, with the Ω given by Eq. 3.2, the points will be restricted to the interior of a sector defined by two phase angles, ϕ_1 and ϕ_2, with $|\phi_2 - \phi_1| < 2\pi$. The width of the sector is given by the fluctuation of $\xi(t)$ around the steady phase Ωt. The frequency in this case can be estimated by a systematic scanning of the stroboscopic plot as a function of τ to find the value τ_S for which the spread of points is restricted to the smallest possible sector of width $|\phi_2 - \phi_1|$, then a new estimate of the frequency, $\Omega_S = 2\pi/\tau_S$, is obtained.

In an efficient implementation of the stroboscopic plot approach, instead of following the dynamics of the system by means of the coordinates $x(t)$ and $y(t)$, this is monitored in a system of cylindrical polar coordinates with the z axis perpendicular to the plane where ϕ is defined, and passing through the center of rotation. The coordinates used are $\rho(t) = \sqrt{x^2(t) + y^2(t)}$, and ϕ defined by Eq. 3.1; and the variables monitored are then $\rho(t)$, and

$$\psi(t) = \phi(t) \pm (2\pi/\tau)\, t, \tag{3.10}$$

with the sign depending on the direction of the rotation. Plots of $\rho(t)$ versus $\psi(t)$ result, in general, in a periodic band of bounded height and unbounded width. The boundedness is because the motion occurs in a bounded region of phase space, and the unboundedness is caused by jumps of 2π of the phase $\phi(t)$ with respect to $\pm(2\pi/\tau)\, t$. However, when $\tau = 2\pi/\Omega$ those jumps do not occur anymore and the stroboscopic plot in the plane $\psi - \rho$ looks like a narrow vertical strip having a width $\Delta\psi < 2\pi$, given as before by the fluctuation $\xi(t)$. The observations in the $\psi - \rho$ and in the $x - y$ plane have to be in mutual agreement, both providing reliable estimates, Ω_S, of the frequency of the chaotic oscillator.

The example of the stroboscopic plot approach to the phase of the Van der Pol–Duffing oscillator is presented in Fig. 3.3. When a series of 2000 consecutive observations of the state of the system are made with a time interval $\tau = 2\pi/\Omega$, the phase space points are restricted to a narrow strip both in the $x-y$ plane, and in the $\psi-\rho$ plane [Figs. 3.3(a, b)] as corresponds to the small fluctuation, $\xi(t)$, around $-\Omega t$, seen in Fig. 3.1(b). When τ is

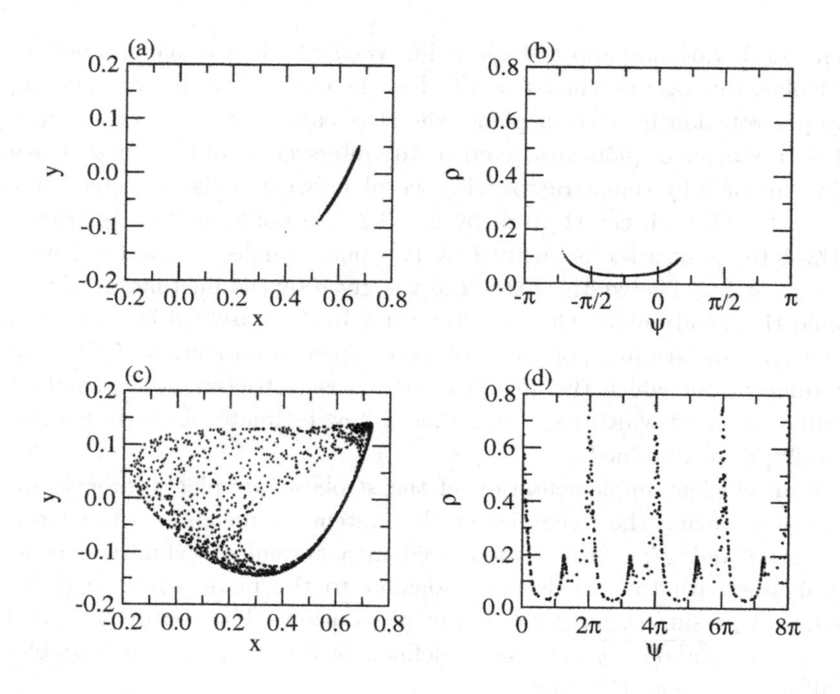

Fig. 3.3 Stroboscopic plots of the Van der Pol-Duffing oscillator taken at time intervals $\tau = 2\pi/\omega$ with $\omega = \Omega = 24.48$ [See Fig. 3.1(d)] projected onto: (a) the $x - y$ plane, and (b) the $\psi - \rho$ plane. Stroboscopic plots taken at time intervals $\tau = 2\pi/\omega$ with $\omega = 1.01 \times \Omega$ projected onto: (c) the plane $x - y$, and (d) the plane $\psi - \rho$. Each dot corresponds to one observation of the state of the oscillator.

slightly increased, the 2000 phase space points observed do spread over the whole attractor region [Fig. 3.3(c)] while the phase difference $\psi(t) = \phi(t) - (2\pi/\tau)t$ growths along a periodic structure [Fig. 3.3(d)] corresponding each period to a 2π jump as $\phi(t)$ moves forward with respect to $(2\pi/\tau)t$.

Finally, in a fourth approach, a frequency of a chaotic oscillator can also be estimated in a simpler way by means of the use of return, or Poincaré maps. In particular, one can obtain the return map for a dynamical observable, $s(t)$, and measure the times, $T_1, T_2, .., T_N$, elapsed between N consecutive maxima, i.e. the return times. The calculation of its average value,

$$\langle T_R \rangle = \lim_{N\to\infty} \left(\frac{\sum_{i=1}^{N} T_i}{N} \right), \tag{3.11}$$

provides the desired frequency $\Omega_R = 2\pi/\langle T_R\rangle$. An analogous definition is possible using the Poincaré map, and the times between consecutive crossings of the Poincaré surface to proceed to Ω_P, as for the return map. Although appealing because of its simplicity, this method should be used with caution because different choices of signals, or Poincaré surfaces, may give different results for the frequency. Also noise or external perturbations may give rise to fake maxima in the return map, or fake crossings of the surface, which will result in unreliable estimates of Ω.

The above discussion assumes that the system dynamics follow a proper rotation, which is a case often found in practice, but not always. For systems with improper rotations the procedure usually followed is to reduce the problem to one or more problems described by simple rotations. In some cases it is possible to find a set of coordinates in which the dynamics follows a proper rotation. For example, this has been done [Pikovsky et al. (1997a)] to study the synchronization of the phase in the Lorenz model [Lorenz (1963)], whose attractor has two centers of rotation in the (x, y, z) space (See Chapter 2); to study this case the use of the new variable $u = \sqrt{x^2 + y^2}$ allows the dynamics to be described as a proper rotation in the plane $u - z$. A more general technique has been proposed [Yalçinkaya and Lai (1997)] in which a scalar chaotic signal, $x(t)$, is decomposed into a superposition of intrinsic modes,

$$x(t) = \sum_{j=1}^{N_m} x_j(t), \tag{3.12}$$

being each mode, $x_j(t)$, a scalar variable whose analytic signal, $\sigma_j(t)$, given by Eq. 3.7 follows a proper rotation. To obtain the first mode the maxima and minima of $x(t)$ are used to fit two spline curves, $x_M(t)$ and $x_m(t)$ respectively, whose arithmetic mean is subtracted from $x(t)$; if the resulting signal is a proper rotation the first mode has been obtained, if not the process is iterated with the new signal until a proper rotation is found. Once the first mode, $x_1(t)$, is found this is subtracted from the original signal and the procedure is repeated again to obtain the second and successive modes until a mode with negligible variation is reached. The above definitions of the phase can then be applied to each mode.

3.1.2 *Characterization of phase synchronization*

A case frequently studied in the literature is when a weak external periodic force is applied to an autonomous chaotic system. This situation can be described by means of a system of differential equations of dimension d of the type

$$\frac{d\mathbf{x}}{dt} = \mathbf{F}(x) + \mathbf{P}(t), \tag{3.13}$$

with $\mathbf{P}(t) = [A_1 \cos(\omega t + \delta_1), A_2 \cos(\omega t + \delta_2), ..., A_d \cos(\omega t + \delta_d)]$ the periodic force of frequency ω that is applied, whose intensity is measured by the amplitudes $A_{j=1,2,\dots,d}$. Under these circumstances it is possible to observe the phenomenon known as phase synchronization. This means that the system remains chaotic but its dynamics become modified in such a way that the phase of the chaotic attractor meets that of the applied force. The occurrence of phase synchronization of a chaotic oscillator to a driving force of frequency ω can be quantitatively formulated, by means of the function

$$\psi(t) = \phi(t) \pm \frac{m}{n}\omega t, \tag{3.14}$$

with m and n whole numbers, as the case when there are two real numbers, ε_1 and ε_2, that verify $\varepsilon_1 < \varepsilon_2$, and $\varepsilon_2 - \varepsilon_1 < 2\pi$, such that

$$\varepsilon_1 < \psi(t) < \varepsilon_2 \tag{3.15}$$

for all t. Using Eq. 3.2, this condition can be rewritten as

$$|n\Omega - m\omega| = 0 \tag{3.16}$$

so that phase synchronization means that the phase of the oscillator always stays close enough to the phase of the force ($m = n = 1$), or to one of its harmonics ($m > n$), or the frequency of the oscillator, Ω, is close to an harmonic of the frequency of the force ($m < n$). If phase synchronization is to be obtained or not, depends on the properties of the force applied: its frequency, ω, amplitude $A_{j=1,2,\dots,d}$, and the angles $\delta_{j=1,2,\dots,d}$.

Because of the different approaches to the phase of an oscillator presented in the previous subsections, phase synchronization can be monitored in several interesting ways. This will be illustrated here by means of an example based on the Rössler model [Rössler (1976)]. The particular form of

Eq. 3.13 used here is

$$dx/dt = -(y+z), \tag{3.17}$$
$$dy/dt = x + 0.2y + A\cos(\omega t), \tag{3.18}$$
$$dz/dt = 0.2 + z(x - 5.7), \tag{3.19}$$

which, for $A = 0$, presents a chaotic attractor which follows a proper rotation. The phase can be studied in the $x - y$ plane, where the dynamical behavior of $\phi(t)$ is quite simple because the projection of the attractor follows a simple counter-clockwise rotation around the origin as shown in Fig. 3.4(a). This makes this system very appropriate to easily visualize phase synchronization. The synchronization condition $\varepsilon_1 < \psi(t) = \phi(t) - \omega t < \varepsilon_2$ with $\omega = 0.167362$ close to the value of Ω at $A = 0$, $\Omega = 0.170778$ will provide neat observations of phase synchronization.

A straightforward approach to the observation of phase synchronization is to follow the dynamics of $\psi(t)$, for increasing values of A as shown in Fig. 3.4(b). For $A = 0$ the difference between the phase of the oscillator and that of the force increases steadily because they are independent, and each corresponds to a different frequency. When the driving is switched on, and becomes intense enough ($A = 0.16$ and $A = 0.18$), time intervals in which $\psi(t)$ verifies a condition of the type given by Eq. 3.15 alternate with small intervals where $\psi(t)$ experiences a fast increase of approximately 2π, which are called phase slips. As the intensity of the driving increases the number of slips decreases, until the system settles down in a well defined state of phase synchronization with the phase difference between the oscillator and the force never greater than 2π. In this example this occurs for $A = 0.19$, where the system oscillates in pace with the force instead of with its own frequency. Moreover, the chaotic attractor [Fig. 3.4(c)] is very much like the unperturbed one. This, however, is not the general case, and under stronger force the attractor can undergo some metamorphosis, and even chaos can be suppressed.

The frequencies obtained from the analysis of Poincaré and return maps through Eq. 3.11 provide an additional procedure to study phase synchronization. These can be obtained for the force of interest, and tested against the condition given by Eq. 3.16. This kind of approach applied to the present example of the Rössler oscillator is displayed in Fig. 3.4(d), where it is shown how the value of Ω_R moves toward ω as the force strength increases. This procedure slightly anticipates the synchronization of the phase

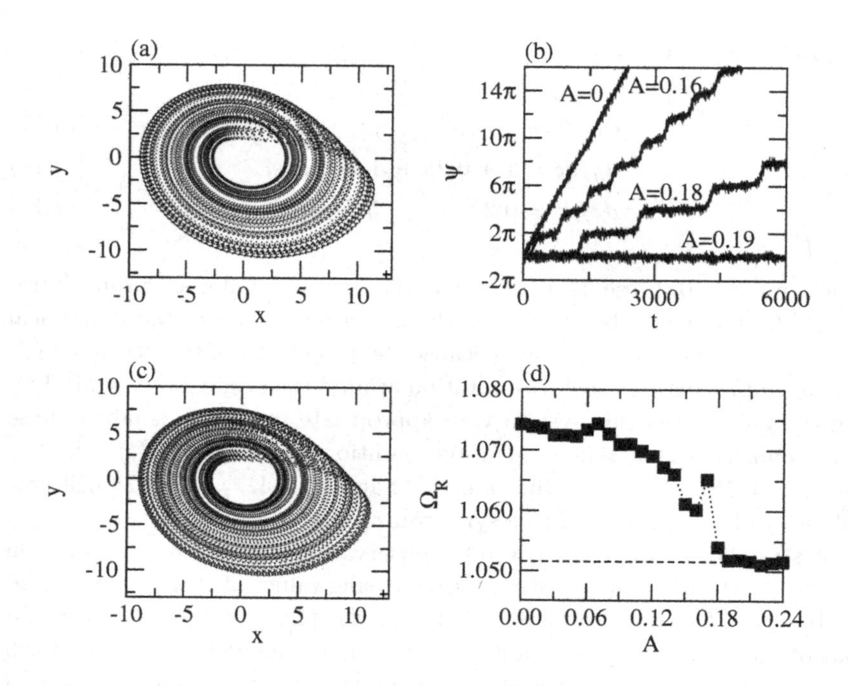

Fig. 3.4 Synchronization of the phase for the periodically driven Rössler attractor. (a) The non perturbed attractor projected onto the $x - y$ plane, and (b) the dynamics of its phase monitored by means of Eq. 3.14 for increasing values of the intensity of the external driving. (c) The projection of the perturbed attractor in conditions of phase synchronization to the force ($A = 0.19$). (d) The synchronization of the phase monitored by means of the frequency of the attractor (squares) computed from the return times [Eq. 3.11] obtained from the maxima of $y(t)$ for increasing values of A. The dashed line signals the frequency of the force, ω.

because $\langle T_R \rangle$ (Eq. 3.11) is practically equal to $2\pi/\omega$ when the number of phase slips is small.

Stroboscopic plots provide an alternative approach to monitor phase synchronization. In this case the values of the system variables are recorded and plotted for fixed time intervals, $\tau = 2\pi/\left(\frac{m}{n}\omega\right)$, given by the frequency of the force. The observations in the $x - y$ plane should display a set of points scattered along the phase space region spanned by the attractor in the desynchronized case and a narrow strip in the synchronized case. The most reliable view made in the $\psi - \rho$ plane should result in a periodic band of bounded height and unbounded width for the desynchronized state, and a narrow vertical strip with a width $\Delta\psi < 2\pi$ for the synchronized state.

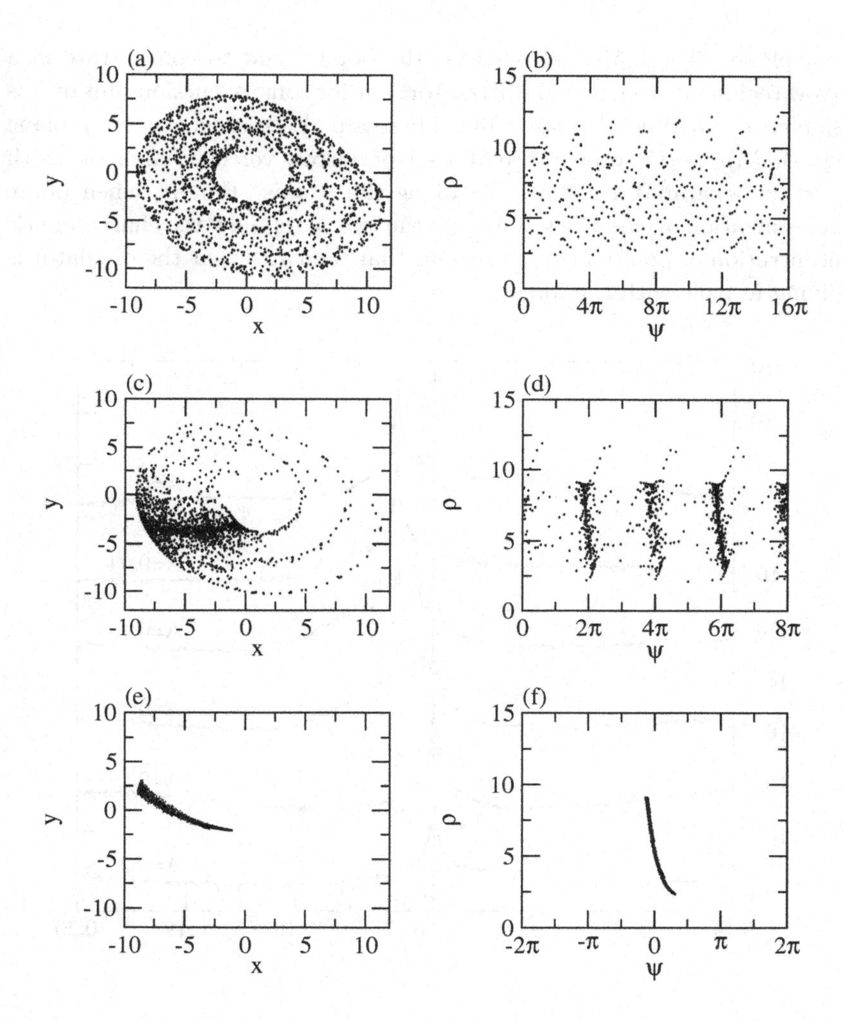

Fig. 3.5 Stroboscopic plot analysis of phase synchronization made in the $x - y$ plane (first column), and in the $\psi - \rho$ plane (second column). The dots correspond to 2000 consecutive observations of the state of the oscillator taken at time intervals $\tau = 2\pi/\omega$. Views of: (a, b) the not synchronized case ($A = 0$), (c, d) the almost synchronized case ($A = 0.18$), and (e, f) the synchronized case ($A = 0.19$).

This view of phase synchronization for the example of the Rössler model is presented in Fig. 3.5. When there is no force applied, Figs. 3.5(a, b) show the characteristic scatter of points onto the $x - y$ and the $\psi - \rho$ planes respectively. When very close to phase synchronization the view in the

$x - y$ plane [Figs. 3.5(c)] shows that the points tend to concentrate in a narrow region on the left of the attractor, but for some excursions out of this region corresponding to phase slips. These are also seen in the $\psi - \rho$ plane [Figs. 3.5(d)] as the appearance of a set of narrow vertical bands of width $\Delta\psi < 2\pi$ separated a distance 2π along the ψ axis. Finally, when phase synchronization occurs [Figs. 3.5(e, f)] the two plots show the characteristic concentration of points in small regions that indicates that the oscillator is evolving in pace with the force.

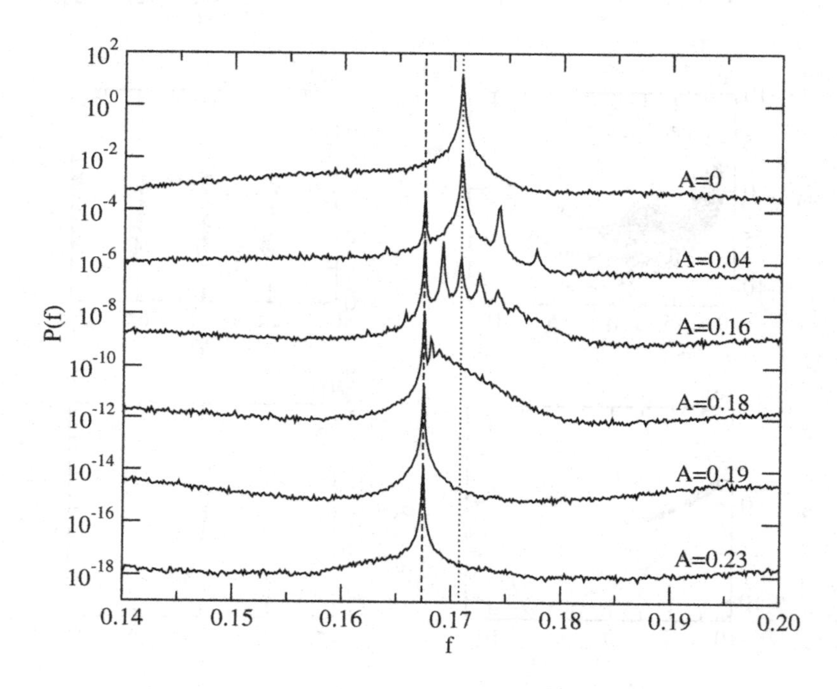

Fig. 3.6 Power spectral density, $P(f)$, of the Rössler attractor for increasing strength of the force, A, in the frequency region around the dominant frequency of the force free attractor (signaled by a dotted line). The frequency of the force is signaled by a dashed line. The plots have been rescaled multiplying $P(f)$ by factors $10^0, 10^{-3}, 10^{-6}, \ldots$ to separate the curves corresponding to each of the force amplitudes, $A = 0, 0.04, \ldots, 0.23$, indicated in the figure.

The observation of the dependence of the power spectral density (Eq. 2.28) of a proper variable of the system on the external force applied provides an additional view of phase synchronization [Pikovsky et al. (1997a)]. Spectral analysis has been useful to provide a qualitative image of the mechanism by which phase synchronization occurs [González-Miranda (2002a); González-Miranda (2002b)]. The main effect of a external periodic force, with a frequency, ω, applied to the oscillator occurs locally in the space of frequencies of the forced system, modifying the importance of the system motions that have frequencies in the neighborhood of ω. This results in a rearrangement of the power spectral values, $P(f)$, around $f = \omega/2\pi$ which enhances the motions that have frequencies closest to $\omega/2\pi$. This may cause two effects: (i) the development of new peaks for f and its harmonics, and (ii) the shift of spectral peaks, that are close to f, towards f. These two mechanisms combined lead to a new structure for the spectral density structure which results in the synchronization of the phase of the oscillator to the external force. The example of the Rössler attractor presented in Fig. 3.6 illustrates this point. When no force is applied the oscillator dynamics has a dominant frequency Ω, which results in a sharp peak for $f_0 \approx \Omega/2\pi$. Even a very small force ($A = 0.04$) results in the appearance of a new peak for $f = \omega/2\pi$. The increase of the force increases the height of this new peak and decreases that of the one at f_0 ($A = 0.16, 0.18$). Finally, the latter completely disappears and the resulting power spectrum has only a dominant peak at f when phase synchronization has been achieved ($A = 0.19, 0.23$).

3.1.3 *Numerical simulations and experimental evidence*

The Rössler model is a simple chaotic flow which, in a wide range of its parameter space, presents a chaotic attractor whose dynamics follow a proper rotation with a phase well defined according to any of the criteria presented in Subsection 3.1.1. Because of this it has been chosen by several investigators to perform theoretical and numerical studies on the basics of phase synchronization. In particular, Pikovsky and collaborators [Pikovsky et al. (1997a); Pikovsky et al. (1997b)] have studied the synchronization of the phase mainly using the criteria given by Eq. 3.15 and by Eq. 3.16 for Rössler models under the parameter values $a = 0.15$, $b = 0.4$, $c = 8.5$, and $a = 0.2$, $b = 1.0$, $c = 9.0$. The forces applied where also different than those presented here: $\mathbf{P}(t) = [A\cos(\omega t), 0, 0]$ for the first set of parameters , and $\mathbf{P}(t) = [yA\cos(\omega t), xA\cos(\omega t + \pi/2), 0]$ for the second. Other authors [Rosa et al. (1998)] have studied a modified version of the Rössler system for parameter values $a = 0.25$, $b = 0.90$, $c = 6.50$ under the force $\mathbf{P}(t) = [0, A\cos(\omega t - \pi/2), 0]$, using the stroboscopic plot approach. Despite these different settings, the overall features of the phenomenology of phase synchronization observed by these authors are the same as those presented in the above subsection. Moreover, they have presented systematic studies of the dependence of the dynamics of the driven oscillator on the force amplitude and frequency, to obtain phase diagrams on the parameter space $\omega - A$ which show the regions of phase synchronization. In all cases, a structure of Arnold tongues similar to that sketched in Fig. 3.7 was obtained. Each tongue has a vertex in $\frac{m}{n}\omega$ and a width that increases with A. Each particular choice of parameters and forces, however, has its own phase diagram; i.e. its particular positions of the tongues vertex, ω, the slopes and shapes of the lines that separate the synchronized from the non synchronized regions, and the ranges of variation of ω and A where the tongue exists.

Besides numerical simulation, several authors have reported experimental evidence of phase synchronization under periodic driving in a variety of systems. Two representative pieces of work will now be discussed.

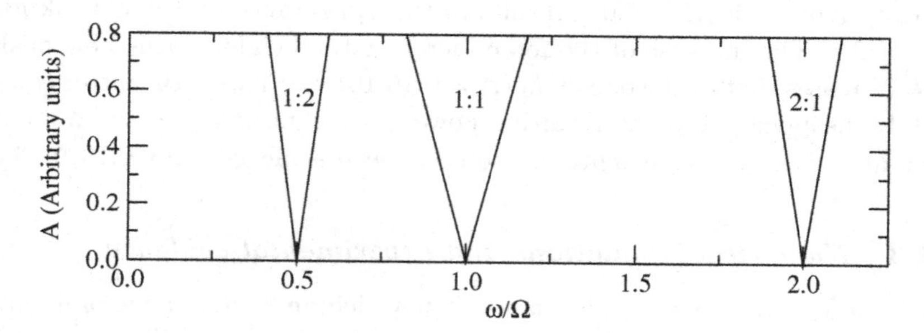

Fig. 3.7 Sketch of the structure of Arnold tongues for a chaotic oscillator of frequency Ω, driven by a periodic force of angular frequency ω, and amplitude A.

The synchronization of the phase in experiments made with a chaotic discharge plasma tube has been reported [Ticos el al. (2000); Rosa et al. (2000)]. The experimental system is a Gleiser tube; i.e. a sealed cylindrical glass capillar tube that contains an unmagnetized plasma and one electrode at each end. These are used to apply a high constant voltage and a low amplitude voltage provided by a sine wave generator, which is the applied force [Fig. 3.8(a)]. The effect of the constant voltage is to produce chaotic oscillations in the intensity of the light, which provides the observable signal here. The sinusoidal voltage is introduced to modify the phase of the oscillations. The system attractor was reconstructed by means of embedding techniques [Abarbanel et al. (1993); Schreiber (1999)]. In this case, phase synchronization was observed and characterized by means of the stroboscopic plot approach [Rosa et al. (1998)]. Plots qualitatively similar to those in Fig. 3.5 and in Fig. 3.7 were reported for the particular events of phase synchronization, and for the phase diagram, respectively.

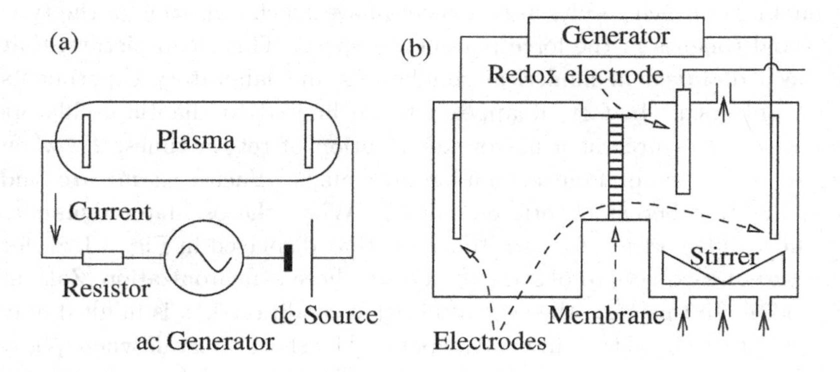

Fig. 3.8 Schematic representation of two of the several experimental devices in which phase synchronization and chaos suppression under weak periodic forces have been observed. (a) Unmagnetized plasma device: an ionized gas (He, Ar, for example) confined between two electrodes is subject to the superposition of a high dc voltage and a weak ac voltage. The measured outputs might be the intensity of the light in the tube (by means of a photo-diode), or the current in the circuit. (b) Continuous stirred electrochemical reaction: the reactants are continuously fed from the bottom of the tank and the products removed from the top (see arrows). The external actions created by a voltage generator are applied to the system by means of metallic electrodes (Pt, Ni, for example). An additional electrode (the redox electrode) provides the redox potential which is a possible output signal.

Synchronization of the phase in a chaotic chemical oscillator has been observed [Kiss and Hudson (2001)] in an experimental setup consisting of a standard electrochemical cell, slowly stirred, and subject to a periodic sinusoidal voltage applied. The system observable was the current of the probe electrode [Fig. 3.8(b)]. By means of a combination of the analytical signal approach [Gabor (1946)] and using intrinsic mode decomposition [Yalçinkaya and Lai (1997)] they obtained proper rotations from measurements of the current of the reference electrode. Synchronization of the phase events similar to those described above for the Rössler model were observed by means of the criteria given by Eqs. 3.15 and 3.16, and by the power spectral density. The phase locked regions found did show Arnold tongues as those sketched in Fig. 3.7.

In this section, perfect manifestations of phase synchronization have been presented and discussed, in which the locking of the phase of the oscillator to that of the force allows a sharp identification by means of several observational techniques. The occurrence of phase synchronization is limited to small applied forces with frequencies being close to have a rational ratio which results in regions of phase synchronization of the type of Arnold tongues in the force parameters space. This clean picture, that has been observed in numerical simulations and laboratory experiments is not universal. In fact, it appears to be limited to chaotic oscillators whose dynamics present a narrow distribution of return times, T_i, when studied by means of Poincaré and return maps. Such systems are said to be highly coherent chaotic oscillators. When the oscillator dynamics present a wide spread of return times (as that displayed in Fig. 3.1(e), for example) one expects to observe imperfect phase synchronization [Zaks et al. (1999)]. In this case, the synchronization condition 3.15 is fulfilled only in a set of finite large time intervals separated by short intervals where phase jumps occur. These phase jumps have been identified [Zaks et al. (1999)] as excursions to phase synchronization states with different locking ratios (Eq. 3.16). Imperfect phase synchronization has been observed in numerical simulations of the Lorenz model [Park et al. (1999)], and later reproduced [Pujol-Peré (2003)] in an experiment made with an electric analog electric circuit designed to mimic the dynamical behavior of the Lorenz model.

3.2 Chaos suppression

A second phenomenon that has been observed in the study of chaotic oscillators driven by a periodic force is the suppression of chaos under weak applied forces. Besides its interest as a phenomenon whose knowledge expands our understanding of the physics on nonlinear systems, it has potential application in the applied sciences and in technology. For one side many systems in the natural sciences are subject to periodic driving; this is the case of the climate system, and of the living organisms on the Earth which are acted on by daily, monthly and yearly perturbations. The understanding of many phenomena in climatology and biology might well require taking into account this effect of chaos suppression. Otherwise in many technical systems chaos may be harmful or undesirable; this occurs in confined plasmas used in thermonuclear fusion, as well as in mechanical, electrical and optical systems. The possibility of eliminating chaos, with little modification of the system is in these cases of great practical interest.

3.2.1 *Weak resonant forces*

Chaos suppression has been mainly studied on periodically forced one-dimensional chaotic oscillators whose dynamics is given by

$$\ddot{x} + f(x, \dot{x}) + \frac{dV(x)}{dx} = A_1 \sin(\omega_1 t), \tag{3.20}$$

where $V(x)$ is a potential function, $f(x, \dot{x})$ is a damping term, and A_1 and ω_1 are the parameters of a external chaos enhancing force. Examples of this kind of system are the driven pendulum (Eq. 1.18) and the Duffing oscillator (Eq. 1.17). Several authors [Lima and Pettini (1990); Braiman and Goldhirsch (1991)] have studied the case when a second harmonic force, $F = A\sin(\omega t)$, is applied to an oscillator which will be in a chaotic state in the absence of this second force. The main result obtained is that chaos may be suppressed for very weak force (i.e., A very small) when the frequency of the force, ω, is one of a set of frequencies, characteristic of the system, which are said to be resonant. It deserves to be stressed that the chaos suppressing force is very weak. Of course, a strong enough periodic force may dominate the system dynamics, turning it periodic; the fact that tiny perturbations may destroy chaos is what is relevant because this means that a very small modification of the system of interest results in a qualitative change of its dynamical behavior.

Two different forms of application of the additional periodic force have been studied. In a straightforward one [Braiman and Goldhirsch (1991)], a second forcing term is applied to the system; i.e.,the force F is simply added to other forces acting on the system, being an additional external force. In this case, a weak perturbation is one for which the amplitude, or strength, of the force A is small compared to the average acceleration of the system, $\ddot{x} = d^2x/dt^2$. A second scheme [Lima and Pettini (1990)] is to perturb one parameter of the system, say p, by forcing it to oscillate around its nominal value, p, as $p[1 + A\sin(\omega t)]$. In this case a small force verifies that $A \ll 1$. This last form of coupling has proven to be easily feasible in experiments performed on different systems.

An example of an externally driven chaotic oscillator is given by the following form of the driven pendulum

$$\ddot{x} + \gamma\dot{x} + \sin x - I = A_1 \sin(\omega_1 t) + A\sin(\beta\omega_1 t), \tag{3.21}$$

which includes an additional constant term, I. The second (chaos suppressing) periodic force depends on two parameters: the strength of the

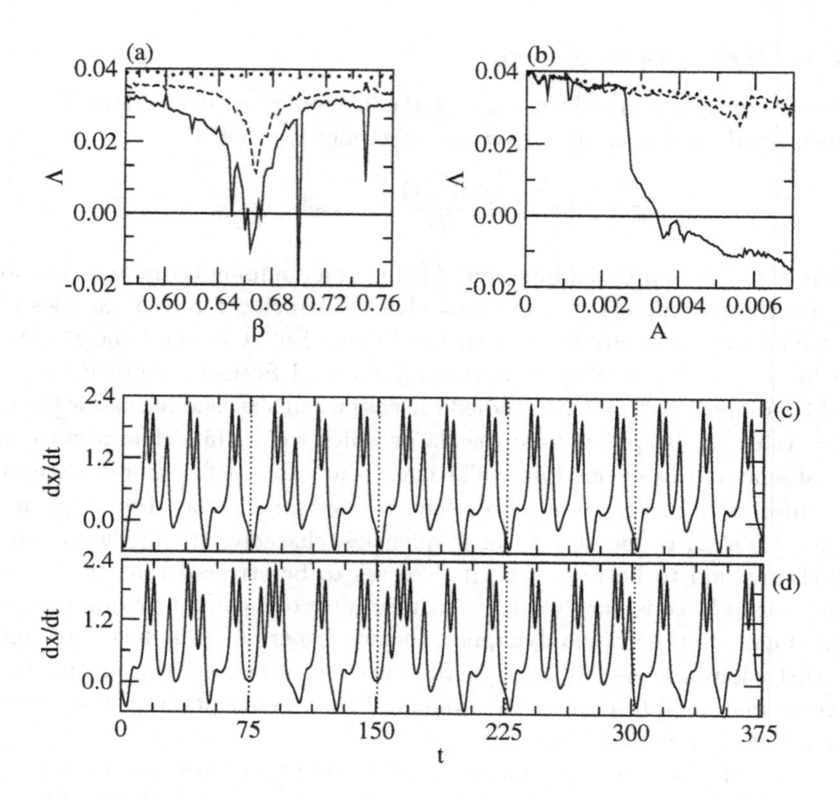

Fig. 3.9 Chaos suppression near the resonant frequency $\beta_r = 2/3$. (a) Transverse Lyapunov exponent as a function of β for $A = 0.001$ (dotted line), $A = 0.003$ (dashed line) , and $A = 0.005$ (continuous line). (b) Transverse Lyapunov exponent as a function of A for $\beta = 0.667$ (continuous line), and two neighbor frequencies: $\beta = 0.570$ (dotted line), and $\beta = 0.770$ (dashed line). (In these two figures the line $\Lambda = 0$ has also been plotted to guide the eye.) Time series plots of dx/dt for $\beta = 0.667$ and (c) $A = 0.007$, (d) $A = 0$. Vertical dotted lines separated by time intervals equal to the estimated period for the periodic case are included to guide the eye.

perturbation, A, and its frequency given in units of ω_1, β. The equations of the same oscillator under parametric perturbation will read

$$\ddot{x} + \gamma\dot{x} + \sin x - I = A_1 \sin(\omega_1 t), \tag{3.22}$$

with

$$\gamma(t) = \gamma[1 + A\sin(\beta\omega_1 t)] \tag{3.23}$$

a parameter that is sinusoidally perturbed by the applied force. This modification of the driven pendulum (Eq. 1.18) happens to be a good description of the electrodynamics of a Josephson junction, a system of interest in the study of superconductivity [Kautz (1996)]. The relevant quantity is $\dot{x}(t)$, which describes the voltage fluctuations in the junction, I is a *dc* current applied to the junction, and $A_1 \sin(\omega_1 t)$ is an *ac* current. For different sets of parameters, and under different driving schemes this is a system that has been used to study, theoretically and computationally, the phenomenon of chaos suppression by means of weak periodic perturbations. For the following examples, Eq. 3.22 will be considered, with the parameter values: $\gamma = 0.7$, $I = 0.905$, $A_1 = 0.4$, and $\omega_1 = 0.2501268$, which according to [Braiman and Goldhirsch (1991)] correspond to an oscillator in a deep chaotic state.

A resonant frequency for which chaos can be suppressed in this system by a weak periodic perturbation is $\beta_r = 2/3 \approx 0.667$. This can be properly monitored by means of the study of the transverse Lyapunov exponent (which is the first no null Lyapunov exponent) as a function of β for increasing values of A. Fig. 3.9(a) shows how the effect of the periodic perturbation is to decrease the degree of chaos, as indicated by the overall decrease of the values of the Lyapunov exponent for all values of β. This, however, is much more intense in the neighborhood of β_r to the extent that the exponent becomes negative in a small interval of values of β around β_r for an amplitude as small as $A = 0.005$. From the study of the Lyapunov exponent for fixed β presented in Fig. 3.9(b), the critical value for which chaos starts to be suppressed for $\beta = 0.667$ can be estimated as $A_c = 0.00345$; while for frequencies 15 percent above and below, chaos is strong and far from being suppressed. Time series of the variable $\dot{x}(t)$ provide additional illustration of this chaos suppression event. In Fig. 3.9(c), corresponding to $\beta = 0.667$ and $A = 0.007$, it is displayed as a somewhat complicated structure that repeats itself with a period $T = 75.36$; while in Fig. 3.9(d), for $A = 0$, no sign of periodicity is observed.

The spectral analysis of chaos suppression presented in Fig. 3.10 is illustrative of the qualitative mechanism by which chaos suppression occurs. This is essentially the same discussed above for phase synchronization. For $A = 0$, the motion is mainly given by a main series of peaks with frequencies $\omega_n = n\omega$ that emerge over a noisy background with a spectral density three orders of magnitude below the height of the main peak [Fig. 3.10(a)]. When the force is turned on, a new series of peaks develops which include the frequency of the force between its harmonics, $\omega_n' = n\beta\omega/2$ [Fig. 3.10(b)]. This is accompanied by a rearrangement of spectral density that results

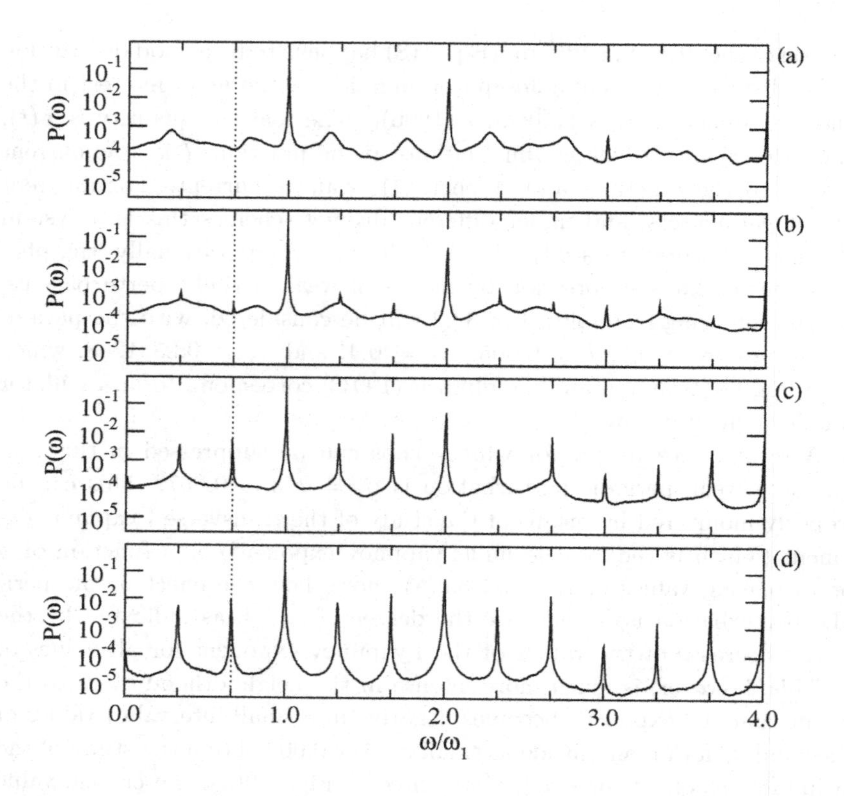

Fig. 3.10 Chaos suppression at the resonant frequency $\beta = 0.667$ studied in Fourier space. Progressive destruction of chaos as the strength of the force increases: (a) $A = 0$, (b) $A = 0.001$, (c) $A = 0.003$, and (d) $A = 0.005$.

in a progressive increase of the height of the new peaks accompanied by a reduction of the level of the background noise [Fig. 3.10(c)] that finally results in the disappearance of noise and then the destruction of chaos [Fig. 3.10(d)].

In all the above studies of chaos suppression it has been assumed that the chaos enhancing force, $F_1 = A_1 \sin(\omega_1 t)$, and the second applied force, $F = A \sin(\beta\omega_1 t)$, are in phase. Some authors [Chacón (1995); Qu et al. (1995); Chacón (1996)] have wondered if a phase difference, δ, between these forces will have some influence on the suppression of chaos. These authors have found that, for values of β which are rational numbers, a forcing of the type $F = A \sin(\beta\omega_1 t + \delta)$ with values of A and β for which there is no chaos suppression when $\delta = 0$, it is possible to find values of δ for which

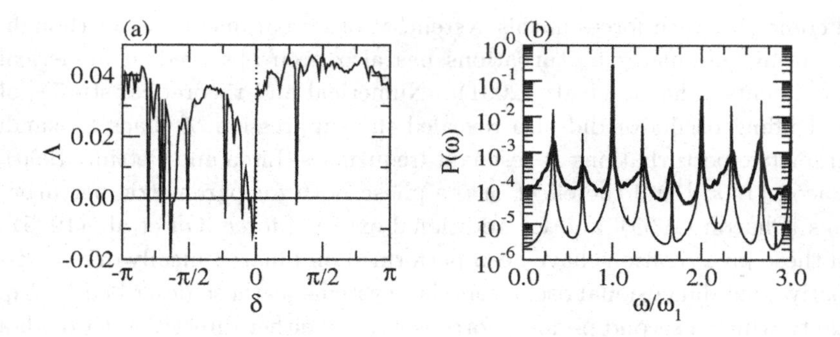

Fig. 3.11 Effect of the phase of the force in the suppression of chaos for $\beta = 1$. Chaos suppression for an appropriate phase as displayed by (a) The transverse Lyapunov exponent for $A = 0.005$, (the line $\Lambda = 0$ has also been plotted to guide the eye) and (b) the power spectral density for $A = 0.005$ and $\delta = 0$ (thick line), or $\delta = -0.2832$ (thin line).

chaos is destroyed, then being the effect of the phase difference to enhance the suppression effect. An example is presented in Fig. 3.11 for the system given by Eq. 3.22 with the same parameter values as before, and a force F with $A = 0.005$, and $\beta = 1$. For $\delta = 0$ the dynamics is chaotic with a positive Lyapunov exponent $\Lambda = 0.0258$; however, when the transverse Lyapunov exponent is studied as a function of δ, one obtains values of the phase difference for which the exponent becomes negative, then signaling the suppression of chaos [Fig. 3.11(a)]. This occurs around $\delta \approx -0.30$ and for $\delta \approx -2\pi/3$. For additional illustration the power spectral densities of the system are displayed for $A = 0.005$, when $\delta = 0$, and when $\delta = -0.2832$ in Fig. 3.11(b): the disappearance of the noisy background, as an effect of the phase difference, is seen there. Similar results could be displayed for the other window of periodic motion.

3.2.2 *Numerical and experimental observations*

The model for the dynamics of a Josephson junction studied here (Eq. 3.22) and the Duffing oscillator (Eq. 1.17) have been the preferred systems to perform numerical and theoretical studies of chaos suppression by several authors. Studies of the Josephson junction using the same parameter values as in the above example but a driving scheme in which an external force term is added to the system, as in Eq. 3.21, instead of a parametric perturbation, did reveal [Braiman and Goldhirsch (1991)] essentially the same phenomenology displayed here in Figs. 3.9 and 3.10. The effect of the phase difference between forces in this system, at other parameter values though, and using parameter perturbations has also been observed in numerical experiments [Chacón et al. (2001)]. Numerical and theoretical studies of the Duffing oscillator did also revealed the suppression of chaos by small parametric perturbations at resonant frequencies [Lima and Pettini (1990); Chacón (1995)] and the effect of the phase both for parametric perturbations [Chacón (1995)] and for additional external force [Qu et al. (1995)]. All these investigations have been performed on biharmonically driven dissipative one-dimensional oscillators, i.e. systems as those described by Eq. 3.20 to which a second periodic force is applied either directly or through a parameter. This is because these kinds of systems are appropriate for the application of a bifurcation theory technique, know as the Melnikov method [Guckenheimer and Holmes (1983)], which allows theoretical predictions of the existence of chaotic motion, and then to substantiate the observations with an analytical treatment [Chacón (1996)].

Beyond driven one-dimensional oscillators, the suppression of chaos under resonant forces applied to chaotic flows (Eq. 2.1) has been observed in numerical simulations. The systems studied are diverse, and two examples drawn from different fields will be discussed here. One case is the FitzHugh–Nagumo neuron model [Scott (1975)], which is a three-dimensional system, that has been studied under parametric perturbation [Rajasekar and Lakshmanan (1994)]; chaos suppression leading to different types of periodic orbits was observed for three different resonant frequencies obeying the relation $1/2 : 1 : 3/2$ for amplitudes of the perturbation as small as 0.010 for the smaller frequencies and 0.030 for the larger one. The other case is a model for a three mode solid state laser [Bracikowski and Roy (1991)], which has dimension six, and has been studied under parametric perturbation of two relevant parameters of the model [Colet and Braiman (1996)], with the observation of chaos suppression for several frequencies and amplitudes; in particular these authors reported quite simple periodic orbits for certain frequency for values of the perturbation as small as 0.006.

This chaos suppression phenomenon has been observed in experiments performed on a variety of systems. One of them, closely related to the analytical and numerical studies of the Duffing oscillator, discussed above, was performed with a simple magneto-mechanical system which realizes a double well pendulum in the laboratory [Fronzoni et al. (1991)]. Periodic perturbations, with amplitudes of the order of 0.10 were applied to the parameters controlling the potential wells. Chaos suppression was observed through time series which display an intermittent behavior with chaotic and regular dynamics interwoven, the length of the regular behavior increased when approaching to certain resonant frequencies.

Besides this, two other experiments that have been performed on devices of the same nature as those described above to discuss the experimental evidence of phase synchronization will be quoted here: an unmagnetized steady-state plasma device [Ding et al. (1994)], and an electrochemical reactor [Guderian et al. (1998)]. Although, the particular experimental devices are different here, the generic descriptions of the experimental setups given above (Fig. 3.8) do still apply.

The experiments on the suppression of chaos on the discharge plasma [Ding et al. (1994)] were made taking the constant voltage applied to the gas to maintain it in a chaotic state as a system parameter to be perturbed by an additional sinusoidal voltage of small amplitude. The suppression of chaos monitored by means of the power spectral density of the signals did show chaos suppression phenomena alike to the one presented in Fig. 3.10: development of peaks at the frequency of the applied force and its harmonics at the cost of the destruction of broad band noise-like structures. Moreover, the maximal Lyapunov exponent was determined using standard embedding techniques [Abarbanel et al. (1993); Schreiber (1999)] to reconstruct the attractor; its study as a function of the frequency of the perturbation, displayed behaviors qualitatively similar to those presented here in Fig. 3.9(a) with narrow regions of chaos destruction around the resonant frequencies. The amplitudes of the perturbations needed to suppress chaos were below 0.025, so that we can properly speak of small perturbations.

The experiment on a chemical oscillator [Guderian et al. (1998)], initially tuned to a chaotic state, was performed by the imposition of a small alternating current whose amplitude was fixed at a small value (0.02 mA), being the frequency the variable used to achieve chaos suppression. Time series of the redox potential, measured with an additional appropriate electrode, and their correspondent power spectral densities were used to monitor the chaotic or periodic state of the system. For appropriate applied frequencies these authors were able to observe periodic orbits of different degrees of complexity depending on the frequency applied.

Chapter 4

Chaotic Oscillators Driven by Chaotic Signals

The most relevant effect observed when a chaotic oscillator is driven by another chaotic oscillator is the synchronization of the chaotic dynamics of the driven system to that of the drive. Two main cases can be distinguished: when the two oscillators are identical or nearly identical, identical synchronization may be observed; and when they are different, a new form of synchronization known as generalized synchronization is expected. Moreover, special forms of synchronization such as marginal synchronization, or anticipated synchronization arise under certain conditions. This chapter will start with a presentation of the different settings that are used to have a chaotic system driven by a chaotic signal. Then, the different forms of synchronization will be presented, with emphasis on techniques of observation, stability of the synchronized state, and conditions of occurrence. A discussion of several numerical simulations and laboratory experiments performed on diverse systems, which include a physiological model, several electric circuits, and lasers of different types will illustrate the properties of synchronization.

4.1 Driving schemes

Since the report by Pecora and Carroll [Pecora and Carroll (1990)] on synchronization of chaotic systems, it has been receiving considerable attention in the literature on chaotic phenomena. These authors found that it is possible to have two identical chaotic systems synchronized under appropriate unidirectional coupling schemes. This is a somehow counter-intuitive result because of the intrinsic instability of chaotic systems that causes exponential divergence of close trajectories. The interest in the study of systems subject to unidirectional coupling has been fed by the expectations of technical and scientific applications. The issues addressed until now include the design of secure communication devices, the study of extended systems and spatiotemporal chaos, and the analyses of meteorological phenomena.

In this chapter, autonomous nonlinear dynamical systems, whose dynamical state is given by a vector $\mathbf{x} \in \mathbb{R}^d$ of d scalar variables, will be considered. Its dynamics is governed by a set of d ordinary nonlinear differential equations

$$\frac{d\mathbf{x}}{dt} = \mathbf{F}(\mathbf{x}). \tag{4.1}$$

It is assumed that the system parameters are such that the steady evolution of the system occurs in a chaotic attractor, $\mathcal{A} \subset \mathbb{R}^d$.

The experimental configuration known as unidirectional coupling will be studied here. In this case, two chaotic oscillators, which for now will be assumed to be identical, are coupled. One of them is called the drive, and it is just a dynamical system whose time evolution is given by Eq. 4.1. The other, known as the response, is a copy of the drive, whose variables $\mathbf{x}'$, in the absence of driving, evolve governed by the same field, $\mathbf{F}(\mathbf{x}')$,

$$\frac{d\mathbf{x}'}{dt} = \mathbf{F}(\mathbf{x}'). \tag{4.2}$$

When the unidirectional drive is established, the response results modified to become a new system with its dynamics governed by

$$\frac{d\mathbf{x}'}{dt} = \mathbf{G}(\mathbf{x}, \mathbf{x}'), \tag{4.3}$$

with $\mathbf{G}(\mathbf{x}, \mathbf{x}')$ verifying the condition

$$\mathbf{G}(\mathbf{x}, \mathbf{x}') = \mathbf{F}(\mathbf{x}), \tag{4.4}$$

for $\mathbf{x}' = \mathbf{x}$. This means that a signal made of the variables of the drive, $\mathbf{x}$, acts on the response, which does not act on the drive. This action becomes null when the two systems follow identical trajectories. This coupling describes a variety of situations that occur in science and technology when the chaotic system of interest is subject to chaotic driving. Two basic particular schemes that realize this type of coupling can be found in the literature: continuous control [Ding and Ott (1994); Kapitaniak (1994)], and replacement of variables [Pecora and Carroll (1991); Cuomo and Oppenheim (1993a)].

The continuous control scheme [Ding and Ott (1994); Kapitaniak (1994)] provides a simple form of unidirectional coupling:

$$\mathbf{G}(\mathbf{x},\mathbf{x}') = \mathbf{F}(\mathbf{x}) + \mathbf{C}\cdot(\mathbf{x} - \mathbf{x}') \tag{4.5}$$

with $\mathbf{C}$ a square matrix of dimension d whose elements are constants, this is multiplied by the vector of differences $(\mathbf{x} - \mathbf{x}')$. The numerical values of the constants $C_{\alpha,\beta}$, that make the elements of the matrix measure the strength of the coupling for each forcing signal, which may be constructed from one, or more, of all the variables of the drive.

As an illustrative example, the case of the Lorenz model [Lorenz (1963)] will be presented now. The drive equations are

$$dx/dt = \sigma\,(y - x)\,, \tag{4.6}$$

$$dy/dt = x\,(r - z) - y, \tag{4.7}$$

$$dz/dt = xy - bz. \tag{4.8}$$

where the system variables are $\mathbf{x} = (x, y, z)$. Many realizations of the continuous control scheme are possible, among them

$$dx'/dt = \sigma\,(y' - x')\,, \tag{4.9}$$

$$dy'/dt = x'\,(r - z') - y' + \gamma\cdot(y - y')\,, \tag{4.10}$$

$$dz'/dt = x'y' - bz', \tag{4.11}$$

is a particular case where the 3×3 matrix $\mathbf{C}$ has only one element that is different to zero: $C_{y,y} = \gamma$; the driving signal is simply y, and the strength of the drive is measured by γ.

In the replacement method [Pecora and Carroll (1991); Cuomo and Oppenheim (1993a)] the drive set up is specified by defining a subset of the variables of the drive, $\mathbf{x}_1 \in \mathbb{R}^m$, with $m \leq d$, and a substitution rule which specifies in which terms in $\mathbf{F}(\mathbf{x}')$ one or more variables in $\mathbf{x}'$ have to be substituted by a variable or combination of variables of $\mathbf{x}_1$. This scheme can be easily understood by means of the example of the Lorenz system. If, as above, the drive is given by Eqs. 4.6–4.8 and the drive signal is $\mathbf{x}_1 = (y)$ a possible choice for the response may be given by [Güémez and Matías (1995)]

$$dx'/dt = \sigma\,(y - x')\,, \tag{4.12}$$

$$dy'/dt = x'\,(r - z') - y', \tag{4.13}$$

$$dz'/dt = x'y' - bz', \tag{4.14}$$

where it is to be noted that the drive signal enters in the first equation but not in the second, and third. Of course, many other realizations are possible.

There is a particular implementation of the replacement method [Pecora and Carroll (1990); Pecora and Carroll (1991)], that has received special attention in the literature, known as subsystem decomposition. In this case, the drive is a nonlinear dynamical system that can be decomposed in two subsystems, described by variables $\mathbf{x}_1 \in \mathbb{R}^m$ and $\mathbf{x}_2 \in \mathbb{R}^n$, with $m + n = d$, being $\mathbf{x} = (\mathbf{x}_1, \mathbf{x}_2)$. Its dynamics is then governed by the equations

$$\frac{d\mathbf{x}_1}{dt} = \mathbf{F}_1(\mathbf{x}_1, \mathbf{x}_2), \tag{4.15}$$

$$\frac{d\mathbf{x}_2}{dt} = \mathbf{F}_2(\mathbf{x}_1, \mathbf{x}_2), \tag{4.16}$$

with $\mathbf{F} = (\mathbf{F}_1, \mathbf{F}_2)$. The evolution of each subsystem is given by the projection of the attractor, $\mathcal{A}$, in the proper subspace: $\mathcal{A}_1 \subset \mathbb{R}^m$ and $\mathcal{A}_2 \subset \mathbb{R}^n$. The response is a copy of the second subsystem, whose variables, $\mathbf{x}_2'$, evolve along

$$\frac{d\mathbf{x}_2'}{dt} = \mathbf{F}_2(\mathbf{x}_1, \mathbf{x}_2'), \tag{4.17}$$

so that it is run by its own variables, $\mathbf{x}_2'$, plus the variables injected from the drive, $\mathbf{x}_1$, which substitute the correspondent variables in the response. For example, if the drive is the Lorenz model (Eqs. 4.6–4.7), a particular implementation could be to choose $\mathbf{x}_2 = (x, y)$ as the second subsystem. The response will then be given by

$$dx'/dt = \sigma\,(y' - x')\,, \tag{4.18}$$

$$dy'/dt = x' \cdot (r - z) - y', \tag{4.19}$$

and the drive signal is simply $\mathbf{x}_1 = (z)$.

4.2 Identical systems

All these drive-response configurations may lead to the synchronization of the dynamics of the response to the dynamics of the drive. This appears surprising when the systems are chaotic because of the intrinsic instability, essential to these systems, whose manifestation is the exponential divergence of close orbits measured by the Lyapunov exponents.

4.2.1 *Synchronization and its stability*

The natural form of synchronization of a response to a drive which is identical is know as identical synchronization. Given a drive system of variables $\mathbf{x}$, with dynamics governed by Eq. 4.1, and an identical response system of variables $\mathbf{x}'$, subject to continuous control or to replacement of variables, it is said that there is identical synchronization of the response to the drive when there are sets of initial condition, $\mathcal{X}_D \subset \mathbb{R}^d$ for the drive and $\mathcal{X}_R \subset \mathbb{R}^d$ for the response, such that for all $\mathbf{x}(0) \in \mathcal{X}_D$ and for all $\mathbf{x}'(0) \in \mathcal{X}_R$

$$\lim_{t \to \infty} |\mathbf{x}'(t) - \mathbf{x}(t)| = 0. \tag{4.20}$$

This definition is still valid in the system decomposition scenario if the initial condition of the response is restricted to $\mathcal{X}_R \subset \mathbb{R}^n$, and the difference $|\mathbf{x}_2'(t) - \mathbf{x}_2(t)|$ is used instead of $|\mathbf{x}'(t) - \mathbf{x}(t)|$.

If the same initial conditions are chosen for drive and response, $\mathbf{x}'(0) = \mathbf{x}(0)$, the two systems will evolve in synchrony in the sense that, $\mathbf{x}'(t)$ will continue being equal to $\mathbf{x}(t)$ for all $t > 0$. The relevant question is then if this synchronized state is asymptotically stable; that is, if small perturbations that make $\delta\mathbf{x} \equiv \mathbf{x}' - \mathbf{x} \neq 0$ will die out exponentially or not. The time evolution of $\delta\mathbf{x}$ is given by

$$\frac{d\,(\delta\mathbf{x})}{dt} = \mathbf{G}(\mathbf{x}, \mathbf{x}') - \mathbf{F}(\mathbf{x}). \tag{4.21}$$

To have asymptotically stable synchronization, the Lyapunov exponents for Eq. 4.21 for $\mathbf{x}' = \mathbf{x}$ must all be negative. These exponents are called conditional Lyapunov exponents because they have to be determined for a certain dynamical state of the drive, $\mathbf{x}\,(t)$, which is given by Eq. 4.1. They are computed from the linearization of Eq. 4.21 around the synchronized state, $\mathbf{x}' = \mathbf{x}$,

$$\frac{d\,(\delta\mathbf{x})}{dt} = \mathbf{J}(\mathbf{x})\delta\mathbf{x}, \tag{4.22}$$

with $\mathbf{J}(\mathbf{x})$ the Jacobian

$$\mathbf{J}(\mathbf{x}) = \left[\frac{\partial \mathbf{G}(\mathbf{x}, \mathbf{x}')}{\partial \mathbf{x}'}\right]_{\mathbf{x}'=\mathbf{x}}, \tag{4.23}$$

of the response vector field for $\mathbf{x}' = \mathbf{x}$. Standard procedures used to compute Lyapunov exponents can then be used to determine the stability of the synchronized state. It deserves to be noted that the application of this stability criterion to a continuous control scenario will lead to a result which is dependent on the matrix of constants $C_{\alpha,\beta}$ that measure the strength of the driving. For the replacement method it provides a yes or no answer.

An alternative to calculate Lyapunov exponents has been proposed [He and Vaidya (1992)], which relies on the use of Lyapunov functions, a standard technique of stability analysis belonging to the qualitative theory of ordinary differential equations [Jordan and Smith (1990)].

These considerations on stability, which assume a response that is a full copy of the drive, can be easily adapted to a subsystem decomposition scenario. The condition for synchronization would be $|\mathbf{x}_2'(t) - \mathbf{x}_2(t)| \to 0$, and the conditional Lyapunov exponents have to be computed from the Jacobian $\mathbf{J}(\mathbf{x}) = [\partial\mathbf{F}_2/\partial\mathbf{x}_2]_{\mathbf{x}_2'=\mathbf{x}_2}$.

All the above will now be illustrated by means of the examples of the Lorenz model introduced in the previous section, using the parameter values $\sigma = 10$, $r = 60$, and $b = 8/3$ for which the Lorenz system is in a chaotic state.

The study of the continuous control configuration described by Eqs. 4.6–4.11 is presented in Fig. 4.1. The stability analysis displayed in Fig. 4.1(a) was made by calculating the conditional Lyapunov exponents as functions of the coupling strength, γ, from the linearized equations for the error $\delta\mathbf{x}$ (Eq. 4.22) by means of standard techniques of calculation of Lyapunov exponents [Wolf et al. (1985)]. This shows that the first and second conditional Lyapunov exponents have both become negative for $\gamma > \gamma_C \approx 3.71$. Because the third exponent is $\lambda_3 < -15$ for all γ, asymptotically stable chaotic synchronization of chaos can be expected for $\gamma > \gamma_C$. This is indeed what is observed by computing $|\delta\mathbf{x}| \equiv |x' - x| + |y' - y| + |z' - z|$ as a function of time from the time evolution of the drive-response systems with initial conditions of the drive chosen in the chaotic attractor, and arbitrary initial conditions for the response. Examples for three different coupling strength intensities are shown in Fig. 4.1(b). For $\gamma > \gamma_C$, the distance between the two systems decreases exponentially in agreement with the synchronization condition: $|\delta\mathbf{x}(t)| \to 0$, when $t \to \infty$. Moreover, the damping effect is larger for larger γ because the absolute value of the leading conditional Lyapunov exponent, which gives the average rate of convergence for close trajectories, increases with γ. This is illustrated in this figure by means of the plot of two additional lines which represent the function $\delta\mathbf{x}(t) = 10 \exp[\lambda_1(\gamma) \cdot t]$ for $\gamma = 5.0$ and for $\gamma = 7.0$. When $\gamma < \gamma_C$ the distance between drive and response oscillates in a range $\delta\mathbf{x} \sim 10$

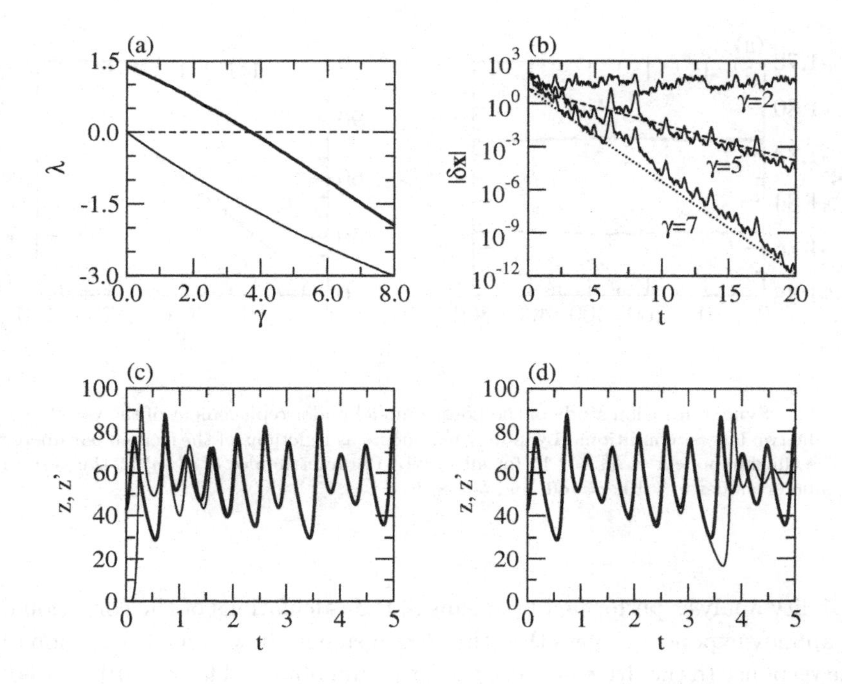

Fig. 4.1 Synchronization study of the Lorenz model under continuous control. (a) The two largest conditional Lyapunov exponents as functions of the coupling strength ($\lambda = 0$ is displayed as a dashed line). (b) Time evolution of $|\delta x|$ for different values of γ as indicated in the labels of the curves; the straight lines show the function $|\delta x\,(t)| = 10\exp(\lambda_1 t)$ for $\lambda_1\,(\gamma = 5) = -0.572$ (dashed line) and $\lambda_1\,(\gamma = 7) = -1.482$ (dotted line). Time series of z (thick line) and z' (thin line) for: (c) $\gamma = 0.50$ and an initial condition with $\delta x\,(0) \sim 10$, and (d) $\gamma = 0.20$ and an initial condition with $\delta x\,(0) < 0.1$.

which is the order of magnitude of the average distance between two points in the chaotic attractor, i.e. this distance stays finite and bound because the drive and the response evolve uncorrelated, but each in its own attractor. For additional illustration Fig. 4.1(c) shows that the time series for the variables z and z', for initial conditions of drive and response separated a distance of the order or the attractor size, are practically identical after 5 cycles for $\gamma > \gamma_C$; while Fig. 4.1(d), for $\gamma < \gamma_C$, shows that z and z' evolve independently after seven cycles, even for nearly identical initial conditions of drive and response ($|\delta \mathbf{x}\,(0)|$ was less than one per cent of the size of the attractor).

As a second example, results for the replacement method obtained when Eqs. 4.12–4.14 were used for the response are presented in Fig. 4.2. The

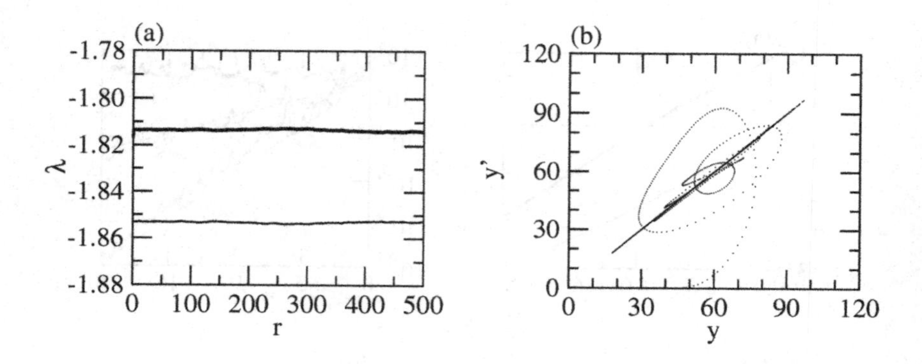

Fig. 4.2 Synchronization study of the Lorenz model under replacement of the variable y. (a) The two largest conditional Lyapunov exponents as functions of the system parameter r (the third exponent is $\lambda_3 \approx -10$ for all r). (b) Parametric plot of $y'(y)$ at the system parameter values $\sigma = 10$, $r = 60$, and $b = 8/3$.

stability analysis performed by means of the calculations of the conditional Lyapunov exponents shows that this arrangement allows synchronization of the response to the drive because the spectrum obtained for $\sigma = 10$, $r = 60$, and $b = 8/3$ is $(-1.813, -1.852, -10.000)$. This is, in fact, approximately independent of the control parameter r, as indicated in Fig. 4.2(a). A particular event of synchronization is displayed in Fig. 4.2(b) by means of a plot of the function $y'(y)$ which is obtained from the parametric equations $y(t)$ and $y'(t)$ that result from the integration of the equations of motion. These parametric plots are frequently found in the literature of this subject, mainly in electronics, where they can be easily obtained in an oscilloscope by means of the input of $y(t)$ as the x-signal and $y'(t)$ as the y-signal. The drive started in a point in the attractor, the response in a state far from the attractor, and the equations of motion were numerically integrated for 20000 time steps. Fig. 4.2(b) shows that, after a transitory evolution, made of about 5 cycles (approximately 500 time steps) the remainder points settle in a segment of the line $y' = y$ which is the characteristic signature of identical synchronization.

Similar examples could had been presented using a scheme of systems decomposition. This will be skipped to avoid being too repetitive, and because this particular form of driving will appear in the next subsection, which deals with special forms of synchronization that are presented by certain classes of chaotic systems.

4.2.2 *Marginal synchronization of chaos*

When the chaotic systems being driven bear some symmetries, they are able to exhibit other interesting synchronization phenomena, beyond the identical synchronization studied above. For spatially symmetric chaotic systems one can observe phenomena such as amplification of chaotic signals, or reproduction of a projection of the drive attractor in a phase space region far from where it is stable [González-Miranda (1996a)]. These are particular forms of a more general synchronization-like behavior known as uniform, or marginal, synchronization of chaos.

A subsystem decomposition scenario like the one described by Eqs. 4.15–4.17 will now be considered. The system Eq. 4.1 is said to be a symmetric system in the subspace $\mathbb{R}^n$ when, given a subsystem of variables $\mathbf{x}_2 \in \mathbb{R}^n$, there is set of coordinate transformations, $T_{\mathbf{P}} : \mathbb{R}^n \longrightarrow \mathbb{R}^n$, characterized by certain parameters $\mathbf{P} = (P_1, P_2, ..., P_n)$, such that the equations of motion are invariant under these transformations. This means that given $\mathbf{x}_2\,(t)$ that verifies Eq. 4.16, the result of the transformation $\mathbf{x}_2^*\,(t) = T_{\mathbf{P}}\,[\mathbf{x}_2\,(t)]$, verifies the same dynamical equation

$$\frac{d\mathbf{x}_2^*}{dt} = \mathbf{F}_2(\mathbf{x}_1, \mathbf{x}_2^*). \tag{4.24}$$

For the observation of marginal synchronization of chaos the set of transformations of coordinates has to be continuous and non-numerable in the following sense. All the transformations, $T_{\mathbf{P}}$, have the same functional form, and are dependent on a set of parameters, $\mathbf{P}$, whose values change continuously in a given interval, $P_i \in [a_i, b_i] \subset \mathbb{R}$, such that $0 \in [a_i, b_i]$ and $T_{\mathbf{P}}(\mathbf{x}_2) \rightarrow \mathbf{I}$, with $\mathbf{I}$ the identity, when $\mathbf{P} \rightarrow \mathbf{0}$. Moreover, given $\mathbf{P} \neq \mathbf{P}'$, it must happen that $|T_{\mathbf{P}}(\mathbf{x}_2) - T_{\mathbf{P}'}(\mathbf{x}_2)| \rightarrow \mathbf{0}$ when $|\mathbf{P} - \mathbf{P}'| \rightarrow \mathbf{0}$, for all $\mathbf{x}_2 \in \mathcal{A}_2$.

Simple and interesting examples are the amplitude transformation

$$T_\alpha(\mathbf{x}_2) \equiv \alpha \cdot \mathbf{x}_2, \tag{4.25}$$

which depends on the parameter $\alpha \in \mathbb{R}_0^+$, and the displacement transformation

$$T_\rho(\mathbf{x}_2) \equiv \mathbf{x}_2 + \rho, \tag{4.26}$$

dependent on the vector parameter $\rho \in \mathbb{R}^n$. The Lorenz equations under y-driving in the scheme of subsystem decomposition defined by Eqs. 4.18–4.19 provide an example of invariance under an amplitude transformation in the $x - y$ plane. The displacement transformation is presented by the circuit of Chua [Matsumoto et al. (1985)]. An inspection of Eqs. 2.15–2.18 shows that the equations for x and z are invariant under the transformation $T_D(z) \equiv z + D$, with $D \in \mathbb{R}$ the parameter defining the transformation. Then, y being the drive signal, the response subsystem whose dynamics is governed by

$$\dot{x}' = \alpha\,[y - x' - f(x')] \tag{4.27}$$

$$\dot{z}' = -\,\beta\, y \tag{4.28}$$

is invariant under a displacement of the attractor in the z direction.

The symmetries described in the above paragraphs have consequences on the dynamical behaviors of these systems when they are under a unidirectional driving scheme. In the absence of external perturbations, if the initial conditions for the response are $\mathbf{x}_2'(0) = \mathbf{x}_2(0)$ with $\mathbf{x}_2\,(0) \in \mathcal{A}_\mathbf{2}$, with the dynamical equations for subsystems $\mathbf{x}_2$ and $\mathbf{x}_2'$ identical it will occur that $|\mathbf{x}_2'(t) - \mathbf{x}_2(t)| = 0$ for all $t > 0$, i.e. the response will evolve in identical synchrony to the drive in the set $\mathcal{A}_\mathbf{2}$. Because of the invariance properties, this implies that for initial conditions $\mathbf{x}_2'(0) = T_\mathbf{P}\,[\mathbf{x}_2(0)]$ with $\mathbf{x}_2\,(0) \in \mathcal{A}_\mathbf{2}$, it will happen that $|\mathbf{x}_2'(t) - T_\mathbf{P}\,[\mathbf{x}_2(t)]| = 0$ for $t > 0$; i.e., the response evolves in synchrony to the drive in the set of points $\mathcal{A}_\mathbf{P} = T_\mathbf{P}\,[\mathcal{A}_\mathbf{2}]$ which is a metamorphosed copy of $\mathcal{A}_\mathbf{2}$. For example, for the case of the amplitude transformation, $T_\alpha(\mathbf{x}_2) \equiv \alpha \cdot \mathbf{x}_2$, the response follows trajectories that reproduce a copy of the drive, amplified (or shrunken) by a factor α. For the displacement transformation, $T_\rho(\mathbf{x}_2) \equiv \mathbf{x}_2 + \rho$, the response follows trajectories that reproduce a copy of the drive, displaced a vector ρ from the region of phase space where the stable attractor lies.

These special synchronization-like behaviors are illustrated in Fig. 4.3 by means of the examples of the Lorenz model and the Chua circuit. For the Lorenz model with the response given by Eq. 4.18–4.19, the projection of a trajectory of the drive onto the $x - y$ plane, and the correspondent trajectory of the response are displayed in Figs. 4.3(a) and 4.3(b) respectively; they show a case when the response portrait is a copy of the drive shrunken by a factor $\alpha \approx 1/2$. The synchrony between the two systems is clearly seen by means of the parametric plots $x'(x)$ and $y'(y)$, which result in straight lines with slopes approximately equal to 1/2 [$x'(x)$ is presented in Fig. 4.3(c)]. So, in this particular case, a response is observed which follows a trajectory which is a small copy of the drive trajectory, and whose time evolution is synchronized with the drive

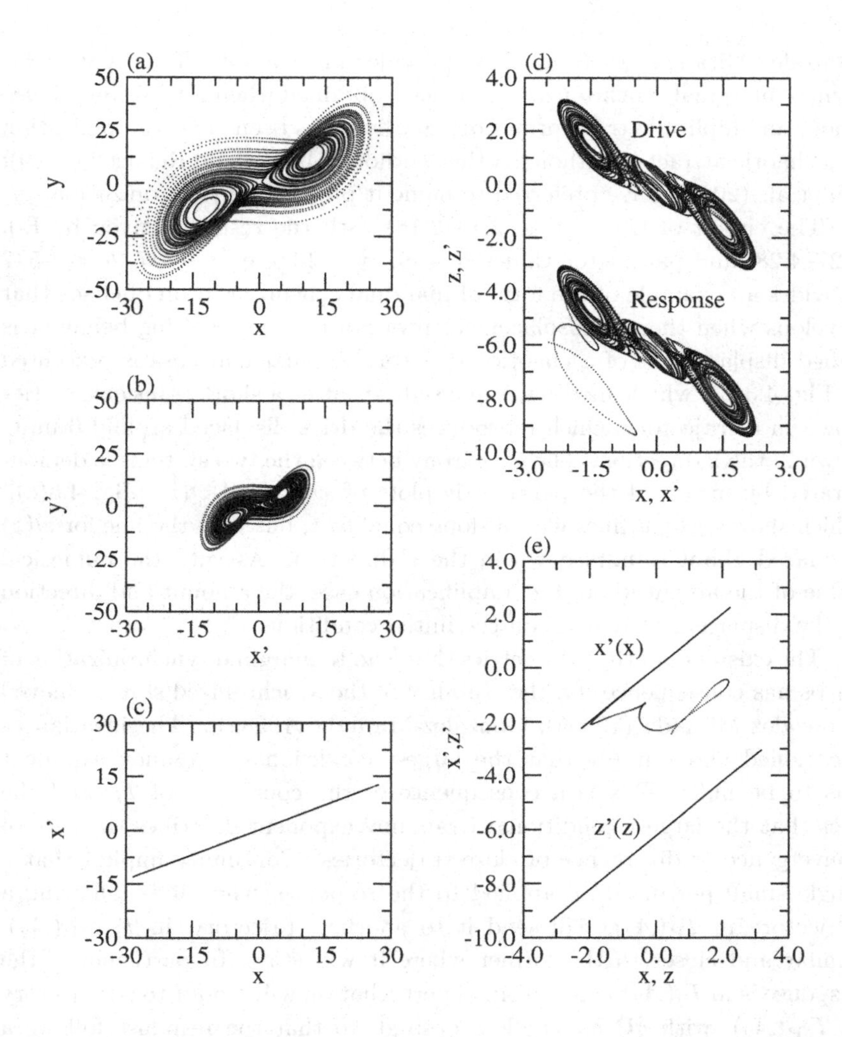

Fig. 4.3 Marginal synchronization of chaos (a–c) of the Lorenz system, and (d, e) of the Chua circuit. Projection of a trajectory of (a) the drive onto the $x-y$ plane, and (b) the response onto the $x'-y'$ plane; (c) parametric plot of $x'(x)$. (d) Projections of trajectories of drive and response in the $x-z$ and $x'-z'$ planes respectively, and (e) parametric plots of $x'(x)$ and $z'(z)$.

in the sense that $|x'(t) - \alpha \cdot x(t)| + |y'(t) - \alpha \cdot y(t)| = 0$. The value of α in this case is approximately equal to $1/2$, but any positive real number is possible, with the particular value of α to be observed dependent on the initial conditions of drive and response [González-Miranda (1996a);

González-Miranda (1998a)]. This provides an example of the particular form of marginal synchronization of chaos obtained when there is invariance under an amplitude transformation that here has been called amplification of a chaotic attractor; although other authors [Mainiei and Rehacek (1999); Xu et al. (2002)] have preferred to name it projective synchronization.

The circuit of Chua (Eqs. 2.15–2.18) with the response given by Eq. 4.27–4.28, and parameter values $\alpha = 9$, $\beta = 14\frac{2}{7}$, $a = -8/7$, $b = -5/7$ provides an example of the form of marginal synchronization of chaos that develops when there is displacement invariance. The resulting behavior is called displacement of a chaotic attractor. A particular case is presented in Fig. 4.3(d), which shows a response that, after a short transient, settles down in a trajectory which reproduces the drive displaced around 6 units down in the z direction. The synchrony between the two systems is demonstrated by means of the parametric plots of $x\prime(x)$ and $z\prime(z)$ [Fig. 4.3(e)], which show straight lines with a slope equal to 1, but with the line for $z\prime(z)$ displaced about 6 units down in the z direction. As with the numerical value of the amplitude in the amplification case, the amount and direction of the displacement depend on the initial conditions.

The existence of the symmetries that lead to marginal synchronization of chaos has consequences on the stability of the synchronized state achieved [González-Miranda (1996a); González-Miranda (1998a)]. The invariances mentioned above imply that the largest conditional Lyapunov exponent has to be null. This is a consequence of the continuity of $T_{\mathbf{P}}$ and the fact that the largest conditional Lyapunov exponent describes the rate of convergence or divergence of close trajectories. Continuity implies that a single small perturbation applied to the response, when it is following a trajectory in $T_{\mathbf{P}}(\mathcal{A}_2)$, will send it to another trajectory, in $T_{\mathbf{P}+\delta\mathbf{P}}(\mathcal{A}_2)$, similar and closer to the former where it will stay. In particular if the response is in $T_{\mathbf{0}}(\mathcal{A}_2) = \mathcal{A}_2$, a small perturbation will send it to a trajectory in $T_{\delta\mathbf{P}}(\mathcal{A}_2)$, with $\delta\mathbf{P}$ as small as desired, so that the response follows a reproduction of the unperturbed trajectory in $\mathcal{A}_2$ as close to it as desired. Therefore, there is neither divergence, $\lambda_1 > 0$, nor convergence, $\lambda_1 < 0$, of close trajectories; and then, $\lambda_1 = 0$. This means that the stability of the response trajectories is not asymptotic, but a special form of stability that is called uniform, or marginal, stability in the literature [Szebehely (1984); Atlee Jackson (1991)]. In this case, a small perturbation applied to the response evolving in a given orbit will send it to another orbit similar and closer to the former, where it will stay. In this sense the type of stability expected here is similar, although somehow stronger [González-Miranda (1996a)], to the type of stability proper of planetary and satellite motion which is known as orbital stability [Szebehely (1984)].

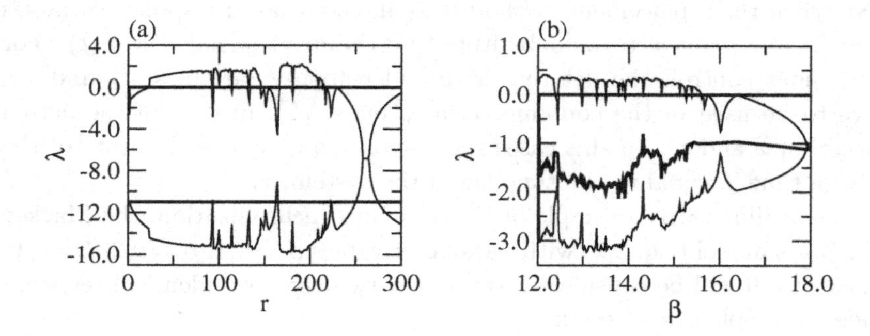

Fig. 4.4 Dependence of the drive Lyapunov exponents (thin lines) and the response conditional Lyapunov exponents (thick lines) on a relevant system parameter for (a) the Lorenz system, and (b) the Chua circuit.

These stability considerations are illustrated in Fig. 4.4, which show the results of calculating the conditional Lyapunov exponents for the response, and the Lyapunov exponents for the drive as functions of a significant system parameter for the above examples of the Lorenz model [Fig. 4.4(a)], and the Chua circuit [Fig. 4.4(b)]. These plots show that the largest conditional Lyapunov exponent, determined by the symmetries of the system, is null and does not change when the system parameters change, despite this changes the degree of chaos of the drive, and eventually the value of the second conditional Lyapunov exponent. This independence on the systems parameters occurs because the value of the exponent is determined by the symmetry of the system which is given by the shape of the equations and not by the parameter values.

4.2.3 *Anticipated synchronization*

Drive-response systems made up of systems described by delayed equations (Eq. 2.8), are able to display a form of synchronization in which the response state, $x'(t)$, synchronizes to a state of the drive that is a time T in the future, $x(t+T)$. This form of synchronization is known as anticipated synchronization [Voss (2000)] because the actual value of the state variable of the response is the same that the correspondent drive state variable will take a time T in the future.

To obtain anticipated synchronization the driving schemes described in Sec. 4.1 have to be modified. For a system with a time delay τ (Eq. 2.8), when the replacement method is used, the delayed response variables $x'_\tau = x'(t-\tau)$ have to be substituted by the drive signal $x = x(t)$. For continuous control, the delayed drive and response variables, x_τ and x'_τ, have to be used in the coupling term, $C(x_\tau - x'_\tau)$, instead of the actual variables, x and x'. In this cases the response anticipates the state of the drive a time T equal to the time lag of the system, τ.

As an illustrative example of this type of synchronization, the Mackey and Glass model (Eq. 2.9) with parameter values $\alpha = 0.1$, $\beta = 0.2$, $C = 10$, and $\tau = 160$ will be considered as the chaotic drive. An identical response under the replacement method:

$$\frac{dx'}{dt} = -\alpha x' + \frac{\beta x}{1 + x^C}, \tag{4.29}$$

will be used for this example. The dynamical behavior obtained in this case is presented in Fig. 4.5. The chaotic evolution of the drive variable $x(t)$ is displayed in Fig. 4.5(a), where a vertical dotted line signals the time $t = \tau$ to provide an insight of the time scale that is presented. In Fig. 4.5(b) the correspondent time series is shown for the response variable which anticipates the state of the drive a time $t = \tau$.

Besides comparison of time series, other procedures have been used to monitor anticipated synchronization. The parametric plot technique used for identical synchronization can be easily adapted just by plotting the delayed variable of the response, $x'_\tau(t)$, against the variable of the drive, $x(t)$. A segment of a straight line with slope equal to 1 is the sign of anticipated synchronization. The application of this procedure to the two time series of this example is presented in Fig. 4.5(c) where the sign of anticipated synchronization is indeed observed.

Some authors [Masoller (2001); Voss (2002)] have resorted to the similarity function , $S(T)$, which quantifies the degree of synchronization between two signals, $x(t)$ and $x'(t)$, as a function of the time shift between them, T. This is defined by means of [Rosenblum (1997)]

$$S^2(T) = \frac{\left\langle [X(t+T) - X'(t)]^2 \right\rangle}{\sqrt{\left\langle [X(t)]^2 \right\rangle \left\langle [X'(t)]^2 \right\rangle}} \tag{4.30}$$

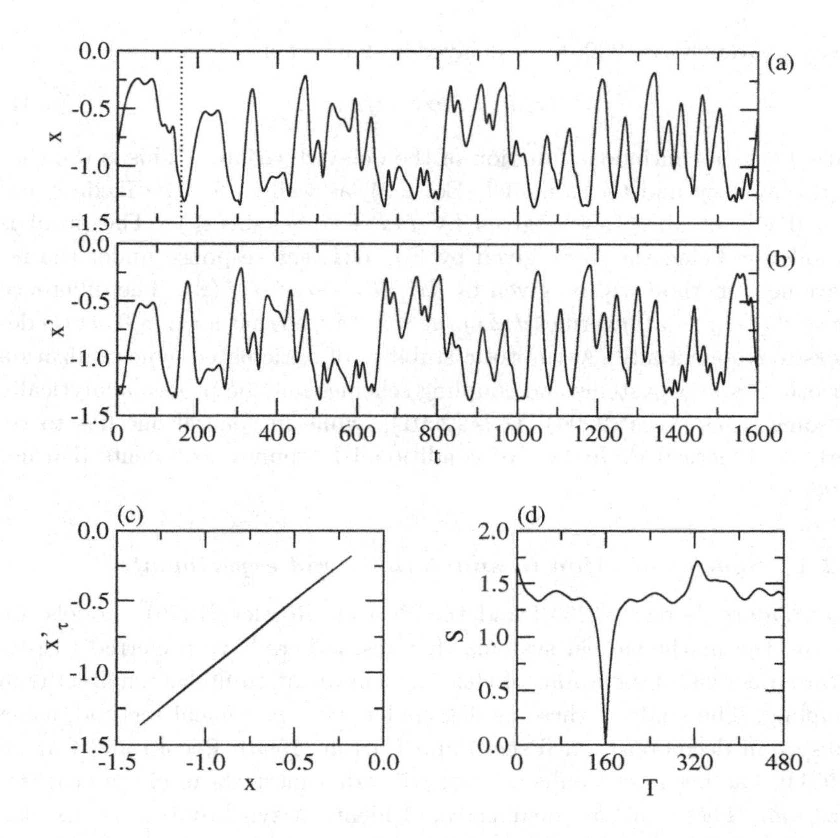

Fig. 4.5 Anticipated synchronization for the Mackey and Glass model, for $\tau = 160$, as displayed by: time series for the state variables of (a) the drive and (b) the response, (c) parametric plot of $x'_\tau(x)$, and (d) the similarity function $S(T)$. The dotted line in (a) signals the time $t = \tau$.

where the angular brackets denote time averages, and $X(t) = x(t) - \langle x \rangle$ is used to have signals with a zero mean. The similarity function takes values of the order of 1 when $X(t+T)$ and $X'(t)$ are not correlated, and becomes null when $X(t+T)$ and $X'(t)$ are identical. Therefore, for the values of $T = T_m$ where $S(T)$ presents a minimum there is synchronization with a time shift equal to T_m. The result of the application of Eq. 4.30 to the two signals from which the time series in Figs. 4.5(a, b) have extracted is displayed in Fig. 4.5(d). This shows, as expected, an acute peak for $T_m = \tau$ which tells that $x'_\tau(t)$ evolves in perfect synchrony with $x = x(t)$.

Anticipated synchronization can be asymptotically stable. This can be easily proven [Voss (2000)] for delayed systems of the form

$$dx/dt = -\alpha x + f(x_\tau), \tag{4.31}$$

with $f(x_\tau)$ a continuous function of the delayed variable. This is the case of the Mackey and Glass model (Eq. 2.9), as well as for the Ikeda equation [Ikeda et al. (1980)], given by $f(x_\tau) = -\beta \sin(x_\tau)$. The proof is as follows: being the drive given by Eq. 4.31, the response under the replacement method will be given by $dx'/dt = -\alpha x' + f(x)$. The difference $\delta x = x'(t-\tau) - x(t)$ verifies $d(\delta x)/dt = -\alpha \delta x$, so that for $\alpha > 0$, $\delta x(t)$ decreases exponentially. Asymptotic stability of anticipated synchronization for other type of systems and coupling schemes may be proven analytically in some cases [Voss (2000); Voss (2001)], while in general one has to resort to numerical evaluation of conditional Lyapunov exponents [Farmer (1982)].

4.2.4 *Synchronization in simulations and experiments*

The Lorenz [Lorenz (1963)] and the Rössler [Rössler (1976)] models are among the mathematical systems that researchers have preferred to perform numerical observations of chaos synchronization under unidirectional coupling. The study of these models under the replacement method, using subsystem decomposition [Pecora and Carroll (1990); Pecora and Carroll (1991)], has been very influential in triggering interests in chaos synchronization. These authors demonstrated identical synchronization, its stability conditions in terms of conditional Lyapunov exponents, its relation to the rate of convergence to the synchronized state, and the structural stability property of the synchronized state. The last refers to the case when drive and response are not identical, but nearly identical; i.e., there are small mismatches between the two systems. Structural stability means that if the identical systems do display synchronization, the nearly identical systems will achieve a state of near synchronization where the distance of the response to the drive, $|\delta \mathbf{x}|$, will not converge to zero, but will remain bound by a quantity much smaller than the size of the attractor. This property is important to warrant the experimental observation of synchronization because of the practical difficulties of having completely identical systems in the real world.

Additional numerical observations of synchronization in the Lorenz model, using the replacement method, have been performed using a response which is a full copy of the drive [Güémez and Matías (1995)]; moreover, these authors illustrated the phenomena of cascade synchronization where a response system acts as a drive for another system, and so on to create complex networks of chaotic oscillators. The Rössler model has also been used [Ding and Ott (1994)] to demonstrate how the continuous control scheme can be used to make negative all the conditional Lyapunov exponents by increasing the coupling strength. Otherwise, the Duffing oscillator (Eq. 1.17) has been used to show that the synchronization under continuous control is robust against external noise [Kapitaniak (1994)]; i.e., if perfect synchronization is achieved in the absence of noise, the addition of a weak noise will not destroy the synchronization, but will lead to a state of near synchronization as the parameter mismatches do. This is also an issue relevant to the practical observation and use of synchronization.

Different forms of marginal synchronization of chaos [González-Miranda (1996a)] have been observed numerically in different systems, and under different forms of the replacement method. The amplification and the displacement of a chaotic attractor have been studied in a system decomposition scheme [González-Miranda (1996a); González-Miranda (1998a)] in the Lorenz model and the Chua circuit, respectively. The study of these phenomena under external noise proved that, for each level of noise, there is a time window in which the synchronization is neatly observed. The width of this window was found to scale with the noise amplitude along a potential law. This means that, despite being marginally stable, this kind of phenomena should be observable in experiments. These and other forms of marginal synchronization of chaos have also been observed in computer simulations of a chemical oscillator [González-Miranda (1999)], in high dimensional systems [Xu et al. (2002)], and in some of the Sprott attractors [Sprott (1994)]; in this last case, using a full copy of the drive for the response [Matías et al. (1997)]. Otherwise, the numerical investigation of marginal synchronization of chaos under a scheme of continuous control suggests that the stability of the synchronized state can be turned from marginal into asymptotic [González-Miranda (1998b); Xu et al. (2001)] by means of a modification of the control mechanism.

Anticipated synchronization [Voss (2000)] was initially observed in the Ikeda model [Ikeda et al. (1980)]. This is a system of delayed equations similar to the Mackey and Glass model studied here, and the phenomenology observed is essentially the same in the two cases. Further numerical studies of the two models [Masoller (2001)] showed that anticipated synchronization is robust against noise and parameter mismatch. The study of chaotic flows, mainly the Rössler and Lorenz models, has been made [Voss

(2000)] under a modified drive-response configuration, given by Eqs. 4.1 and 4.3, with delayed variables, $\mathbf{x}'_\tau = \mathbf{x}'(t-\tau)$ used in the coupling setup. The anticipation times, τ, for which synchronization is observed in these cases use to be small compared to the characteristic times of the systems considered. However, it can be significantly increased by means of a cascade of drive-response systems [Voss (2001)]. This confers to anticipated synchronization the character of a quite general phenomenon.

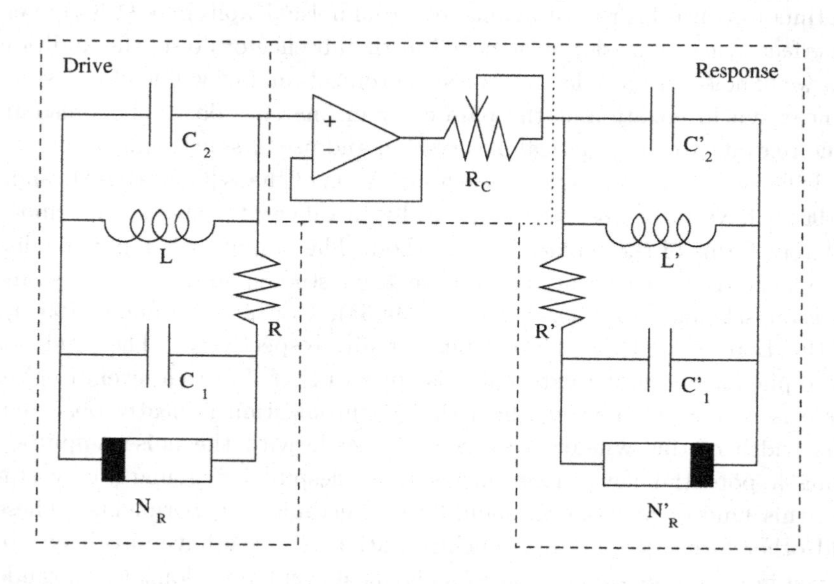

Fig. 4.6 Example of an experimental setup to study the synchronization of chaos in directionally coupled chaotic circuits. The circuit of Chua (See Chapter 2) is considered in this example. The drive circuit is on the left and the response circuit on the right. This last has been modified with a coupling circuit, which includes an operational amplifier, to create a continuous control coupling in the y variable, with the intensity of the coupling controlled by the variable resistance R_C.

Several electric circuits have been used to demonstrate that the phenomena described above in mathematical and computational terms are observable in real-world experiments. Different schemes of unidirectional coupling can be implemented by applying a signal (usually a voltage) from some point of the drive circuit to the appropriate point of the response through an analog circuit designed to model the desired driving scheme. This can be an operational amplifier set up as an impedance converter to simply inject a signal [Carroll and Pecora (1991);

Voss (2002)] on the response, or a more complex circuit performing subtraction and addition of signals to realize continuous control [del Rio et al. (1994)]. This type of setup is illustrated in Fig. 4.6 by means of the particular example of the circuit of Chua [Matsumoto et al. (1985)] driven by continuous control in an arrangement which realizes the following equations for the response

$$dx'/dt = \alpha\,[y' - x' - f(x')], \tag{4.32}$$

$$dy'/dt = x' - y' + z' + (y - y')/R_C, \tag{4.33}$$

$$dz'/dt = -\beta y'. \tag{4.34}$$

Many experiments on synchronization of chaos under a scheme of variable replacement have been performed. An experimental realization of chaos synchronization on a modified version of a hysteretic circuit [Newcomb and Sathyan (1983)] using a system decomposition scheme [Carroll and Pecora (1991)] showed synchronization by means of oscilloscope images of parametric plots of voltages measured at identical points of drive and response which displayed lines with an angle of 45° similar to those shown in Fig. 4.2(b). In the same way, chaotic synchronization under the variable replacement method on a response system which is a full copy of the drive has been demonstrated in an analog electronic circuit implementation of the Lorenz model [Cuomo and Oppenheim (1993b)]. Phenomena of identical synchronization of chaos, using continuous control, have been reported [del Rio et al. (1994)], in an electric circuit [Afraimovich et al. (1986)] whose design realizes the dynamics of a system of mathematical interest [Arneodo et al. (1981)] in the laboratory.

Moreover, there has been an experimental observation of marginal synchronization of chaos [Chua et al. (1992)] using a realization of the system decomposition scheme on the circuit of Chua which implements the set up described by Eqs. 4.27–4.28. In this case a sustained displacement of the attractor similar to that described in Fig. 4.3(e) was observed.

Anticipated synchronization of chaos has also been observed in coupled electric circuits [Voss (2002)], in a system where the nonlinear function $f(x)$ (Eq. 4.31) was a third degree polynomial. This system, which presents a dynamics that resembles the Mackey and Glass model, displayed phenomena of anticipated synchronization similar to those shown in Fig. 4.5.

Synchronization of chaos has also been reported in different types of lasers. The basic setup for this kind of experiment is presented in Fig. 4.7. A part of the light of a drive laser is separated in two parts by a beam

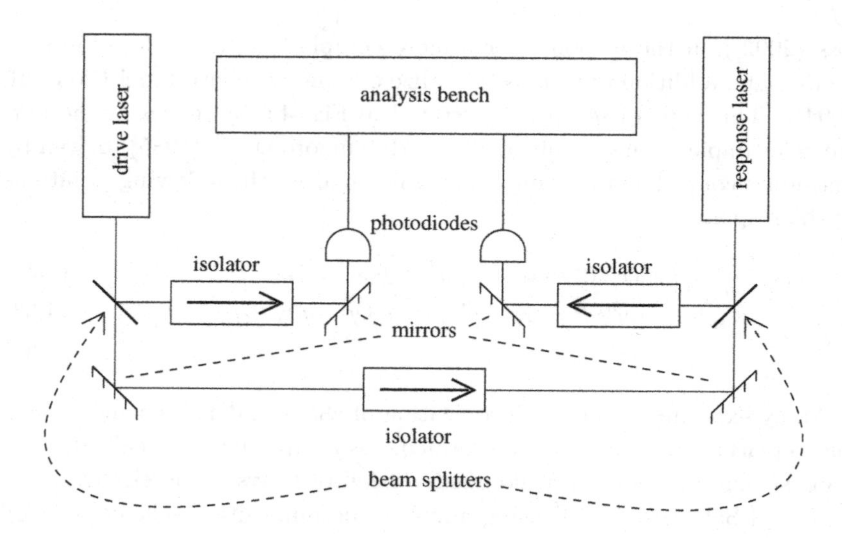

Fig. 4.7 Scheme of a generic experimental setup to study the synchronization of chaos in directionally coupled chaotic lasers. No particular type of laser is considered, but in any case they are prepared in such a way that they output a part of its light which is used both for analysis, and to perform the coupling.

splitter. One is sent to a photodiode which produces an electrical signal to be analyzed, and the second is sent to the response laser by means of an optical insulator to achieve one-directional coupling. This signal is injected on the response laser, whose output is sent to a second photodiode and then to the analysis bench. Usually, this contains an oscilloscope, which allows the observation of the laser signals as time series, as well as parametric plots. In this type of experiment the amount of light that is sent to the response is the control parameter which measures the intensity of the coupling. Although the details, such as the kind of laser or other devices used may change from one experiment to another, the scheme in Fig. 4.7, provides an overall idea of the nature of the observations that have been made. Reports of chaos synchronization has been made in a variety of lasers which include: CO_2 lasers [Sugawara et al. (1994)], external-cavity laser diodes [Sivaprakasam and Shore (1999)], and solid state lasers [Uchida et al. (1999)]. In these cases, the synchronization was monitored by comparison of the time series of the outputs of drive and response, and by parametric plots. Anticipated synchronization has also been observed in chaotic semiconductor lasers unidirectionally coupled [Tang and Liu (2003)], the experimental setup is basically the same presented in Fig. 4.7, with the main modification made in the lasers themselves which were converted in time delayed systems by means of an optoelectronic feed-back loop added to bring back part of the light to the laser with a certain time delay.

4.3 Non-Identical systems

Identical synchronization has been presented in the above section as a form of synchronization characteristic of identical chaotic systems. When the response is definitely different than the drive, identical synchronization is impossible; however, in this case it is possible to observe generalized synchronization, which includes identical synchronization as a particular case.

4.3.1 *Generalized synchronization*

In the above study of synchronization under unidirectional coupling it has been assumed that the response is an identical, or nearly identical, copy of the drive (or of a part of it). However, it is easy to imagine interesting practical situations where the response is different from the drive. The first case is when the response system parameters are very different to the corresponding parameters of the drive. This occurs when the drive and the response are systems of the same nature, i.e. the two have the same physical structure, but they bear individual differences which might be large. For example, the drive could be an electric circuit having a certain design, or disposition of its elements (resistors, capacitors, ...), and the response a copy of this circuit having the same design but with one or more particular elements being clearly different (having different values for resistances, capacities, ...). The second case is when the nature of the two systems is different; i.e. the systems are structurally different. Following the example of the electric circuit, the drive and the response could be both electric circuits, but having a different design, i.e. different elements arranged in a different manner; or even, the drive could still be an electric circuit controlling a system of completely different nature, such as an optical system, or a biological tissue.

The generalization of the driving schemes described in Sec. 4.1 to the case of a response that is different to the drive is straightforward. There is still the drive, Eq. 4.1, whose variables are given by the vector $\mathbf{x} \in \mathbb{R}^d$, and whose nature and dynamics is modeled by the vector field $\mathbf{F}(\mathbf{x})$. The response, is a new system, described by new variables $\mathbf{y} \in \mathbb{R}^r$, which when free evolves governed by a different field, $\mathbf{\Phi}(\mathbf{y})$,

$$\frac{d\mathbf{y}}{dt} = \mathbf{\Phi}(\mathbf{y}). \tag{4.35}$$

When the driving is set up, the response is modified to become a new system with dynamics governed by

$$\frac{d\mathbf{y}}{dt} = \mathbf{\Gamma}(\mathbf{x}, \mathbf{y}), \tag{4.36}$$

which can be realized, for example, by means of a continuous control scheme, $\mathbf{C}\cdot(\mathbf{x} - \mathbf{y})$, or by some variable substitution rule. However, in general, $\mathbf{\Gamma}(\mathbf{x}, \mathbf{y})$ does not allow a condition similar to Eq. 4.4 because the different nature of drive and response usually does not allow the condition $\mathbf{x}(t) = \mathbf{y}(t)$ for all t.

This, however, does not mean that synchronization of the response to the drive is impossible when the response is not identical, nor nearly identical, to the drive. In fact, it is possible to find in practice cases when the dynamics of the response is completely determined by the dynamics of the drive, i.e. a given trajectory of the drive allows the unambiguous prediction of the trajectory of the response. This phenomenon is known as generalized synchronization [Rulkov et al. (1995)], and it is defined as the case when there is a set of initial conditions $\mathcal{X} \subset \mathbb{R}^d$ for the drive, and $\mathcal{Y} \subset \mathbb{R}^r$ for the response such that for all $\mathbf{x}(0) \in \mathcal{X}$, and for all $\mathbf{y}(0) \in \mathcal{Y}$,

$$\lim_{t\to\infty} |\mathbf{y}(t) - \phi[\mathbf{x}(t)]| = 0, \tag{4.37}$$

with $\phi[\mathbf{x}(t)]$ a functional relation which determines the phase space trajectory of the response, $\mathbf{y}(t)$, from the trajectory of the drive, $\mathbf{x}(t)$. It has to be noted that the functional $\phi[\mathbf{x}(t)]$ has to be the same for all initial conditions in $\mathcal{X} \times \mathcal{Y}$. For the particular case when $\phi[\mathbf{x}(t)]$ is the identity, $\phi[\mathbf{x}(t)] = \mathbf{x}(t)$, the definition of identical synchronization (Eq. 4.20) is recovered.

The detection of generalized synchronization is not an easy task in many cases, and certain criteria and algorithms have been developed for this aim. There are three main approaches to the observation of generalized synchronization: analysis of conditional Lyapunov exponents[Kocarev and Parlitz (1996)], the auxiliary system approach [Abarbanel et al. (1996)], and the use of statistical estimations of predictability [Rulkov et al. (1995); Schiff et al. (1996); Breakspear and Terry (2002)].

Being a generalization of identical synchronization, generalized synchronization has to be asymptotically stable [Kocarev and Parlitz (1996)]. This means that for any trajectory $\mathbf{x}(t)$ of the drive in its stable attractor, given by Eq. 4.1, and for the correspondent trajectory $\mathbf{y}(t)$ of the response, given by Eq. 4.36, all response trajectories resulting from a small enough perturbation, $\mathbf{y}(t) + \delta\mathbf{y}(t)$, should converge exponentially to $\mathbf{y}(t)$. This implies that the Lyapunov exponents of the linearized equations for the time evolution of $\delta\mathbf{y}(t)$,

$$\frac{d(\delta\mathbf{y})}{dt} = \left[\frac{\partial\mathbf{\Gamma}(\mathbf{x},\mathbf{y})}{\partial\mathbf{y}}\right]_{\mathbf{x}=\mathbf{x}(t),\mathbf{y}=\mathbf{y}(t)} \delta\mathbf{y}, \tag{4.38}$$

have to all be negative. Therefore, a necessary and sufficient condition for generalized synchronization, is that all of the r conditional Lyapunov exponents of the response have to bee negative. These can be computed by means of the same technique used to study the stability of identical synchronization. Lyapunov functions[He and Vaidya (1992)] can also been used to this aim [Kocarev and Parlitz (1996)].

The auxiliary system approach[Abarbanel et al. (1996)] is a test for generalized synchronization which is based on the predictability property of synchronization: if two identical copies of the response that start with different initial conditions within the synchronization basin of attraction, are driven by the same drive system, the final state achieved after transitories have died off has to be the same for the two copies. Therefore, the auxiliary system approach resorts to the use of an additional response system which is identical to the response, and independent of it. It is described by the variables $\mathbf{y}'\in\mathbb{R}^r$, governed by the field, $\mathbf{\Phi}(\mathbf{y}')$, in the absence of drive, and by the field $\mathbf{\Gamma}(\mathbf{x},\mathbf{y}')$, when the drive is switched on. If the dynamics of the $d+2r$ system given by Eq. 4.1, Eq. 4.36, and $d\mathbf{y}'/dt = \mathbf{\Gamma}(\mathbf{x},\mathbf{y}')$, collapses to a manifold that verifies $\mathbf{y}(t) = \mathbf{y}'(t)$; then, there is generalized synchronization of the drive to the response in the sense that there is a transformation, $\mathbf{y}(t) = \phi[\mathbf{x}(t)]$, that gives the dynamics of the response from the dynamics of the drive, with $\phi[\mathbf{x}(t)]$ a locally continuous, point to point, non time dependent transformation [Abarbanel et al. (1996)]. Parametric plots of the auxiliary system versus the response provide an easy test for generalized synchronization.

An example of a statistical test for detecting generalized synchronization is the calculation of the mutual false nearest neighbors parameter, μ [Rulkov et al. (1995)]. In particular, this, as well as others that have been proposed [Schiff et al. (1996); Breakspear and Terry (2002)], is a test of local neighborliness between two time series that assumes that $\phi\left[\mathbf{x}\left(t\right)\right]$ is smooth. This implies that given a time series of the vectors of phase space variables of the two oscillators, $\mathbf{x}(t_k)$ and $\mathbf{y}(t_k)$ with $k = 1, 2, ..., N$, two close states in $\mathbf{y}(t_k)$ have to correspond to two close states in $\mathbf{x}(t_k)$. Given the drive state $\mathbf{x}(t_k)$ and its correspondent response state $\mathbf{y}(t_k)$, each will have a nearest neighbor in the time series; being $\mathbf{x}(t_k^{(x)})$ and $\mathbf{y}(t_k^{(y)})$ which occur at times $t_k^{(x)}$ and $t_k^{(y)}$, respectively. With these elements, the following parameter is defined:

$$\mu = \frac{1}{N}\sum_{k=1}^{N} \frac{\left\|\mathbf{y}(t_k) - \mathbf{y}(t_k^{(x)})\right\|}{\left\|\mathbf{x}(t_k) - \mathbf{x}(t_k^{(x)})\right\|} \frac{\left\|\mathbf{x}(t_k) - \mathbf{x}(t_k^{(y)})\right\|}{\left\|\mathbf{y}(t_k) - \mathbf{y}(t_k^{(y)})\right\|}, \tag{4.39}$$

where the symbol $\|\cdot\|$ indicates the Cartesian distance. From Eq. 4.37 it follows that this parameter has to take a value of the order of 1, if there is generalized synchronization [Rulkov et al. (1995)]. If not, μ has to be a large number whose magnitude is comparable to the product of the size of the attractors divided by the product of the distances between nearest neighbors in the time series, $\mathbf{x}(t_k)$ and $\mathbf{y}(t_k)$. This provides a procedure to detect and quantify generalized synchronization.

It is to be noted that the assumption of smoothness of $\phi\left[\mathbf{x}\left(t\right)\right]$ is made, which is not guaranteed in all cases [Pyragas (1996)]. When $\phi\left[\mathbf{x}\left(t\right)\right]$ is not smooth problems of detectability appear that can be worked out in some cases [Pyragas (1996); Rulkov and Afraimovich (2003)].

A standard example of generalized synchronization is given by the dynamics of a Lorenz model [Lorenz (1963)] driven by a Rössler model [Rössler (1976)]. The equation of motion for the drive are

$$dx_1/dt = -\left(x_2 + x_3\right) \tag{4.40}$$

$$dx_2/dt = x_1 + 0.2 \cdot x_2 \tag{4.41}$$

$$dx_3/dt = 0.2 + x_3(x_1 - 5.7) \tag{4.42}$$

and for the response

$$dy_1/dt = 10\left(y_2 - y_1\right), \tag{4.43}$$

$$dy_2/dt = y_1\left(60 - y_3\right) - y_2 + C\cdot\left(x_2 - y_2\right), \tag{4.44}$$

$$dy_3/dt = y_1 y_2 - \frac{8}{3} y_3, \tag{4.45}$$

which is driven by continuous control, being C a measure of the strength of the coupling. A test of Generalized synchronization made by means of

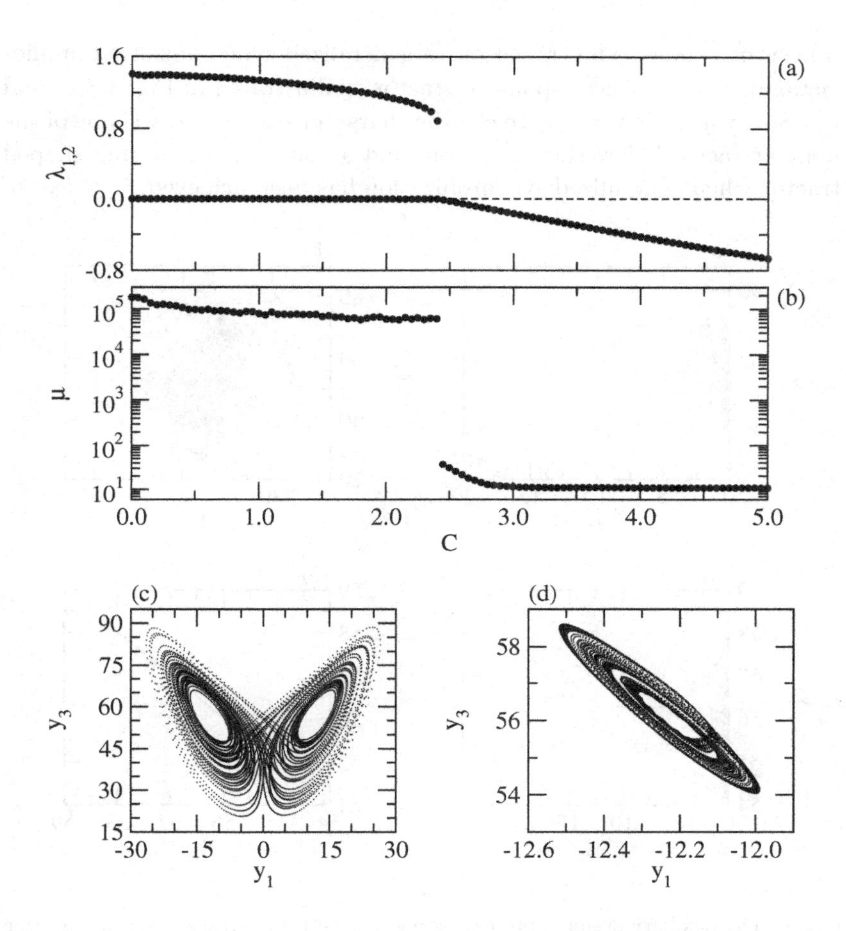

Fig. 4.8 Tests of generalized synchronization applied to a Lorenz model driven by a Rössler model: (a) the two largest conditional Lyapunov exponents (the third exponent lies below $\lambda = -15$), and (b) the mutual false nearest neighbor parameter, both as functions of the coupling strength. Plots of the projection of the response trajectory: (c) for $C = 2$ and (d) for $C = 3$.

the calculation of conditional Lyapunov exponents as functions of the coupling strength, presented in Fig. 4.8(a) shows that the response is unstable until a synchronization threshold around $C \approx 2.45$, were the positive and null exponent have become both equal and negative. This is in agreement with a second test based on the calculation of the mutual false nearest neighbor parameter, presented in Fig. 4.8(b), which shows how this parameter drops four orders of magnitude once the synchronization threshold has been overcome. The transition to generalized synchronization implies a metamorphosis of the response attractor as illustrated in Fig. 4.8(c) and Fig. 4.8(d) which show respectively, the characteristic butterfly shape of the Lorenz attractor below the transition, and a completely new ring shaped attractor when generalized synchronization has been achieved.

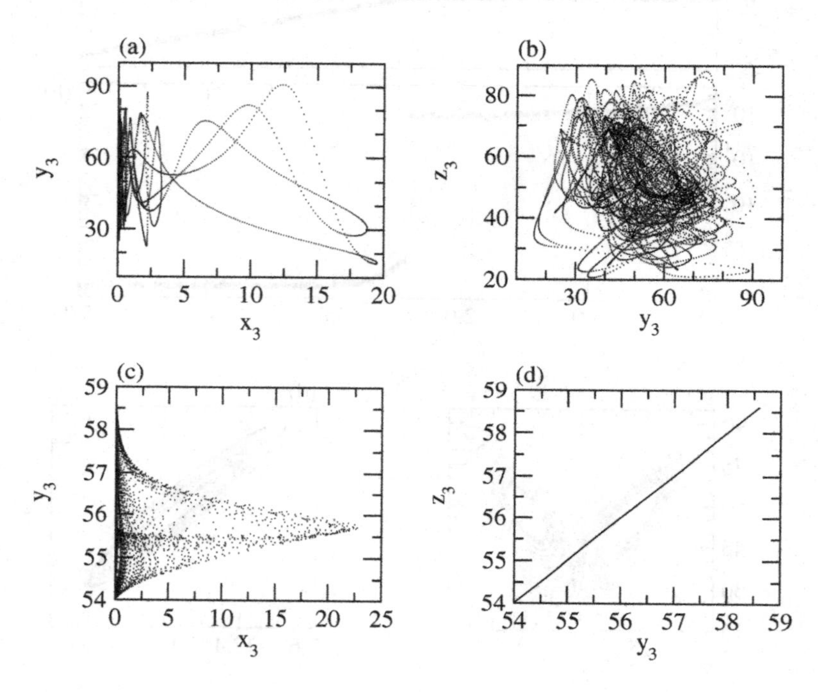

Fig. 4.9 The auxiliary system approach applied to detect generalized synchronization in a system made of a Lorenz oscillator driven by a Rössler oscillator. Parametric plots below the synchronization threshold ($C = 2$) for the third system variable: (a) the response versus the drive, and (b) the auxiliary system versus the response. (c) and (d) are the same as (a) and (b) respectively, obtained in a regime of generalized synchronization ($C = 3$).

The application of the auxiliary system approach to the detection of generalized synchronization for the drive-response given by Eqs. 4.40–4.45 is presented in Fig. 4.9. Below the synchronization threshold the parametric plots of response versus the drive [Fig. 4.9(a)] and of the auxiliary system versus the response [Fig. 4.9(b)] result both in cloudy distributions of points which are indications of a lack of identical and generalized synchronization. Above the synchronization threshold the parametric plot of the response versus the drive is still cloudy [Fig. 4.9(c)], while the plot of the auxiliary system versus the response [Fig. 4.9(d)] shows that the cloud of points has collapsed to a straight line segment with slope equal to one, which is the sign of generalized synchronization, all in accordance with what has been presented in Fig. 4.8.

Despite generalized synchronization being introduced as a form of synchronization proper of non identical systems, it is possible to find reports in the literature of the observation of generalized synchronization phenomena when drive and response are systems with identical individual dynamics. Therefore, non-identity of the systems is not a necessary condition for generalized synchronization.

This occurs in symmetric chaotic systems whose invariant property leaves the equations of motion unchanged in certain subspace, and are driven by means of the variables of the other subspace [González-Miranda (1996b)]. For example, a system which is invariant under rotations of π radians in a certain plane, has two possible synchronization states: one characterized by the synchronization condition $D_0(t) \equiv |\mathbf{w}'(t) - \mathbf{w}(t)| \to 0$, and another given by $D_\pi(t) \equiv |\mathbf{w}'(t) + \mathbf{w}(t)| \to 0$. The first is a case of identical synchronization , while the second is generalized synchronization given by the simple functional relation $\phi[\mathbf{w}(t)] = -\mathbf{w}(t)$.

A particular example is the following chaotic flow [González-Miranda (1996b)]

$$\dot{x}_1 = x_2 + 3.2\ \sin(1.4\ x_2), \tag{4.46}$$

$$\dot{x}_2 = -x_2 - (x_3 - R)\ x_1, \tag{4.47}$$

$$\dot{x}_3 = x_1^2 - x_3, \tag{4.48}$$

which is invariant under a rotation of 180° in the plane $x_1 - x_2$ (i.e. under a change of sign in x_1 and x_2). A possible response system that illustrates the effect of this symmetry is

$$\dot{y}_1 = y_2 + 3.2\ \sin(1.4\ y_2), \tag{4.49}$$

$$\dot{y}_2 = -y_2 - (x_3 - R)\ y_1, \tag{4.50}$$

$$\dot{y}_3 = y_1^2 - y_3, \tag{4.51}$$

which is able to reproduce the drive rotated 180° for initial conditions chosen in the appropriate basin of attraction. Some authors call this dynamical behavior antiphase synchronization [Cao and Lai (1998)], or antisynchronization [Wedekind and Parlitz (2002)]. It could be argued that this form of generalized synchronization is quite trivial as it is essentially the identical synchronization studied in Sec. 4.2, with the only difference being the coexistence of symmetric attractors [Cao and Lai (1998)].

The particular system given by Eqs. 4.46–4.51, besides identical synchronization and the above form of generalized synchronization due to symmetry, is able to present a non-trivial form of generalized synchronization for initial conditions in wide regions of the initial conditions six-dimensional space [González-Miranda (1996b); González-Miranda (2002c)]. This is a complicated form of generalized synchronization for which the functional relation $\phi [\mathbf{x}(t)]$ cannot be determined easily; although it can be detected by means of the above techniques. The origin of this generalized synchronization state, can be traced by a qualitative analysis of the global dynamics of the coupled six-dimensional system. Because there is a basin of initial conditions leading to identical synchronization, the response phase space points whose coordinate values are close to coordinate values of points of the drive attractor will follow trajectories that lead them to reproduce the dynamics of the drive. In other words, the drive trajectory is an attractor for the response. In the present case, the introduction of the coupling leads to the appearance of a new set of stationary points which happen to be attractive in the response subspace, and compete with the attraction to the drive trajectory. If y_3 is sufficiently large, the balance between these two attractions results in the non-trivial generalized synchronization attractor [González-Miranda (2002c)].

These synchronization behaviors are illustrated in Fig. 4.10 which presents results of a numerical study of Eqs. 4.46–4.51 made for $R = 5.2$. The symmetry of the attractor in the $x_1 - x_2$ plane is clear from the plot of the projection of the drive attractor presented in Fig. 4.10(a). The dynamics of the response, depending on the initial conditions, has two main outputs available: one which results in the reproduction of the drive attractor [inner loop in Fig. 4.10(b)], and the other which results in trajectories which display an enlarged and metamorphosed image of the drive attractor trajectories [outer loop in Fig. 4.10(b)]. Because of the inversion symmetry around the origin, displayed in Figs. 4.10(a, b), each of these two qualitatively distinct behaviors of the response allows oscillations of the response in phase or antiphase with the drive. When the drive attractor is reproduced there is identical synchronization for oscillations in phase, and antiphase synchronization otherwise. The latter is illustrated in Fig. 4.10(c) which shows how the minima of the time series of the response correspond with the maxima of the time series of the drive. In the other case there is a non-

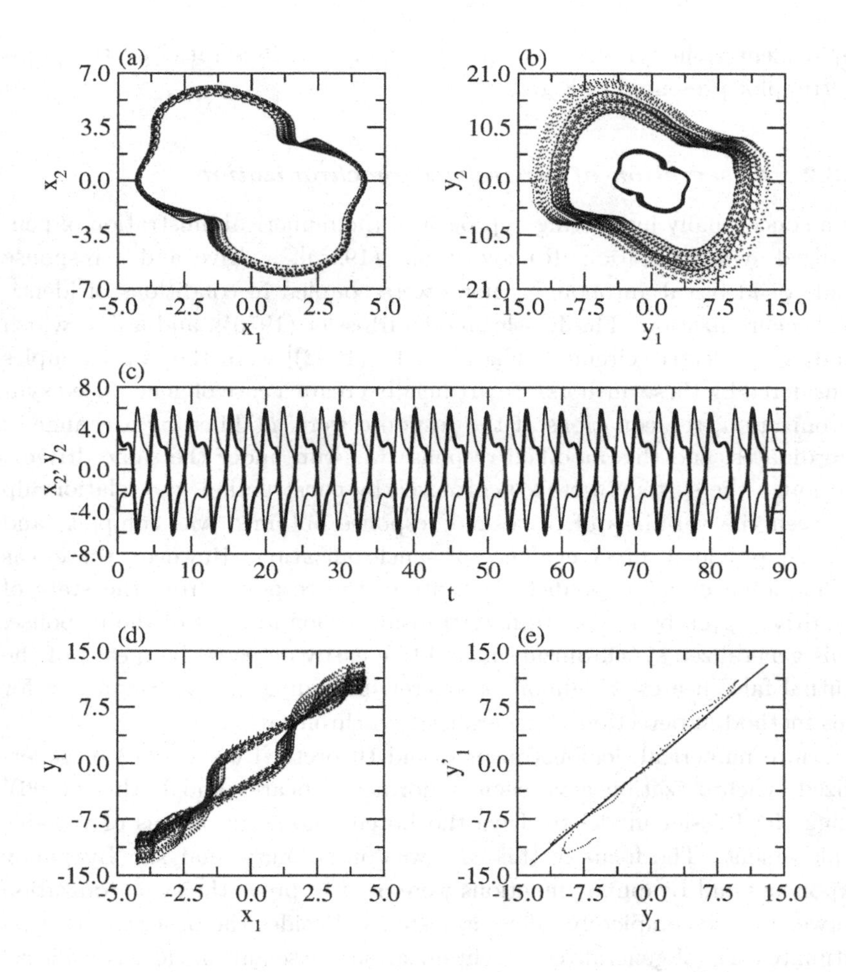

Fig. 4.10 Synchronization of the system given by Eqs. 4.46–4.51. (a) Projection of the drive onto the $x_1 - x_2$ plane. (b) Projections of the response for initial conditions leading to antiphase synchronization (inner loop) and to generalized synchronization (outer loop). (c) Time series from the antiphase synchronization attractor for the drive (thin line), and for the response (thick line). Parametric plots from the generalized synchronization attractor for (d) the response versus the drive, and (e) the auxiliary system versus the response.

trivial form of generalized synchronization. When this state is tested for identical synchronization by means of parametric plots it gives a negative [Fig. 4.10(d)], but when a second copy of the response, (y_1', y_2', y_3'), started at different initial conditions than the response is acted by the same drive, rapidly enters in synchrony with the response as illustrated by the parametric plot presented in Fig. 4.10(e).

4.3.2 *Observation of generalized synchronization*

In a conceptually interesting approach to the numerical illustration of generalized synchronization [Rulkov et al. (1995)], a drive and a response made of identical autonomous flows were coupled in conditions of identical synchronization. The Rössler model [Rössler (1976)], and a flow which models an electric circuit [Rulkov et al. (1992)] were the two examples considered by these authors. To artificially create a case of generalized synchronization, the equations of the response were modified by a change of coordinates, and this modified response was run under the same driving. Although the systems were synchronized by construction, the relationship between the variables of drive and response obtained was complex, and did not provided direct evidence of synchronization. However, there was a functional $\phi[x]$, to predict the state of the response from the state of the drive, given by the coordinate transformation applied to the response. This generalized synchronization could be easily detected by means of the mutual false nearest neighbor parameter, providing in this way a test for this method of detection of generalized synchronization.

More numerical demonstrations (and theoretical discussion) of generalized synchronization have been performed [Kocarev and Parlitz (1996)] using the Rössler model to drive the Lorenz model by means of variable replacement. The focus in this case was on stability analysis: Lyapunov exponents and Lyapunov functions were used to prove the synchronization between these completely different systems. Besides the observation of legitimate cases of generalized synchronization, these authors also considered the case of marginal synchronization of chaos [González-Miranda (1996a)]. In this case, they noted that although there is a functional relation between the variables of drive and response (e.g., amplification or displacement of the attractor as shown in Subsection 4.2.2), marginal synchronization is not a particular manifestation of generalized synchronization because the largest Lyapunov exponent being equal to zero there is no asymptotic stability. In other words: different initial conditions of the response lead to different synchronized states. Further study of the case when drive and response are different, and the largest conditional Lyapunov exponent is null has lead [Krawiecki and Sukiennicki (2000)] to a definition of the concept of generalized marginal synchronization of chaos, which was illustrated with various numerical simulations made with the Lorenz model and the Chua circuit.

Computer simulations of symmetric chaotic systems displaying antiphase synchronization of chaos, as well as more complicated synchronization behaviors associated to other symmetries different to the inversion around a point have been performed [González-Miranda (1996b)] with emphasis on the possible instabilities that result from the coexistence of different synchronization states in the same phase space. Other authors [Cao and Lai (1998)] have numerically demonstrated the antiphase synchronization of chaos in the case of hyperchaotic oscillators.

Generalized synchronization has been observed in experiments made with structurally identical electric circuits whose chaotic dynamics had been previously studied in detail [Rulkov et al. (1992)]. These circuits worked in different chaotic attractors because the numerical values of the circuits components where different. The unidirectional coupling was of a continuous control type. In a set of experiments [Rulkov et al. (1995)] the mutual false nearest neighbor parameter was used to detect the development of generalized synchronization with increasing coupling strength. In a second series of experiments [Abarbanel et al. (1996)], a third auxiliary electric circuit, which was an identical copy of the response, and subject to the same driving showed the development of generalized synchronization, for a coupling strength large enough, as the collapse of a parametric plot to a straight line.

The case when the drive and response circuits are structurally different has also been studied experimentally [Kittel et al. (1998)], by means of the driving of the Shirinki circuit [Shinriki et al. (1981)] by an electronic analog implementation of the oscillator of Mackey and Glass [Mackey and Glass (1977)] by continuous control. Lyapunov exponents and various statistical tests showed the emergence of generalized synchronization with the increase of the strength of the coupling.

There have also been reports of the observation of generalized synchronization of chaos in experiments made with lasers. In one of them [Tang et al. (1998)], an NH_3 laser was pumped on with a previously recorded signal coming from a function generator whose amplitude could be varied to control the intensity of the drive. The signal resulted either from a numerical simulation of the Lorenz equations [Lorenz (1963)], or a laser signal operating in a different chaotic state. A modified version of the auxiliary system approach was used in which different evolutions of the same response system, driven by the same pre-recorded signal were compared in a parametric plot. Again, generalized synchronization of the laser intensity to the driving signal was observed for strong enough driving.

The form of synchronization of chaos between identical chaotic systems called above antiphase synchronization has been observed in directionally coupled semiconductor lasers [Wedekind and Parlitz (2002)]. These authors presented time series which showed how, for appropriate initial conditions, when the intensity of the drive drops, the intensity of the response jumps up and vice versa, which is the sign of antiphase synchronization.

Chapter 5

Perturbing Chaotic Systems to Control Chaos

The idea of chaos control is based on the fact that chaotic attractors have a skeleton made of an infinite number of unstable periodic orbits which are visited for short periods of time by a phase space point which follows a trajectory in the attractor. The aim of chaos control is to stabilize a previously chosen unstable periodic orbit by means of small perturbations applied to the system, so the chaotic dynamics is substituted by a periodic one chosen at will among the several available. This makes chaotic systems very interesting because they allow different uses, without performing structural changes, and employing a minimal external input. Two basic techniques have been developed, and will be presented here: the OGY method, and the Pyragas method. Moreover, some relevant experimental realizations, which include magneto-mechanical systems, fluid dynamics, chemical reactions, and biological systems will be presented.

5.1 The OGY method

Control of chaos, understood as stabilization of unstable periodic orbits embedded in a chaotic attractor, [Ott et al. (1990); Shinbrot et al. (1993)] has been achieved in a variety of systems by means of a technique, or some modifications of it, that has a solid theoretical background on the analysis of the properties of the phase space in the neighborhood of an unstable periodic orbit [Ott et al. (1990)].

5.1.1 *Stabilization of unstable periodic orbits*

It has been demonstrated, theoretically [Ott et al. (1990)] and experimentally [Ditto et al. (1990)], that it is possible to stabilize previously chosen unstable periodic orbits of a chaotic system by means of the application of appropriate small perturbations on a single system parameter, p. To do this the dynamics of the system is followed in a Poincaré surface of section (defined in Sec. 2.2), and the perturbation applied to the control parameter, p, is dependent on the actual state of the system. The control mechanism is discrete in the sense that in a time continuous system the perturbation is applied once per orbit.

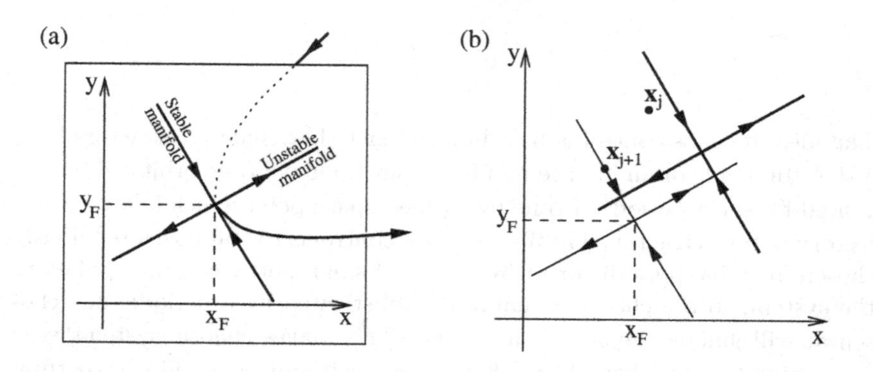

Fig. 5.1 (a) A surface of section (square) is intercepted from behind by an unstable periodic orbit (dotted line) in the point $\mathbf{x}_F$, which has coordinates x_F and y_F, measured on the surface. The motion of the Poincaré map around $\mathbf{x}_F$ is characterized by a stable manifold (thick line with arrows in) and an unstable manifold (thick line with arrows out) which intersect in $\mathbf{x}_F$. (b) When a small change, carefully chosen, is made on a system parameter, the fixed point and its manifold (thin lines) move to a near region in the surface of section (thick lines); then, an iterate from the state $\mathbf{x}_j$ can be sent to $\mathbf{x}_{j+1}$ on the stable manifold of $\mathbf{x}_F$.

The idea of this method, known as the OGY method, is illustrated in Fig. 5.1. It is assumed that a Poincaré map $\mathbf{x}_{j+1} = \mathbf{F}(\mathbf{x}_j; p)$ is constructed from the intersection of the trajectories of a continuous chaotic flow, which describes the dynamics of the system to be controlled, with a surface of section. An unstable periodic orbit, embedded in the flow, intersects from behind the surface of section, represented by a square, in the point $\mathbf{x}_F$ [Fig. 5.1(a)]. A coordinate system $x - y$ constructed on the surface of section is used to label the point $\mathbf{x}_F$ by means of its coordinates, x_F and y_F. This is a fixed point of the Poincaré map, $\mathbf{x}_{j+1} = \mathbf{F}(\mathbf{x}_j; p)$, to which the dynamics of the chaotic flow is reduced onto the surface of section. Because the periodic orbit is unstable, there must be a direction on the surface of section, associated with a positive Lyapunov exponent of the unstable periodic orbit. Along this direction there is exponential divergence of the trajectories generated by the map from points in the neighborhood of $\mathbf{x}_F$; this is called the unstable manifold, and is represented by a line with arrows pointing out of the fixed point in Fig. 5.1(a). Because the periodic orbit is bound inside the chaotic attractor, there must be a negative Lyapunov exponent, which induces a direction in the surface of section along which trajectories are exponentially attracted to the fixed point; this is the stable manifold, which in Fig. 5.1(a) is represented by a line with arrows pointing towards the fixed point. In general the motion generated by the Poincaré map around $\mathbf{x}_F$ is unstable because the stable components of a trajectory die off exponentially, while its unstable components grow at an exponential rate too. Therefore, $\mathbf{x}_F$ is an unstable fixed point of the map, $\mathbf{x}_{j+1} = \mathbf{F}(\mathbf{x}_j; p)$, which is characterized by its stable and unstable manifolds.

The chaos control method by means of small parametric perturbations is sketched in Fig. 5.1(b) where the surface of section and its coordinate system are plotted. The fixed point, $\mathbf{x}_F$, and its stable and unstable manifolds, which appear as thin arrowed lines, correspond to a certain value, p_0, of the control parameter, p. If this parameter changes to $p_0 + \delta p$, with δp small enough, the essential nature of the system dynamics will be preserved, and the main effect of the change will be to move the fixed point and its characteristic manifolds to a new close position which in Fig. 5.1(b) are represented by thick lines. The unstable periodic orbit can be stabilized when it is possible to define a control law such that a value of δp can be assigned to each point $\mathbf{x}_j$ in the neighborhood of $\mathbf{x}_F$ with the property that $\mathbf{x}_{j+1} = \mathbf{F}(\mathbf{x}_j; p_0 + \delta p)$ falls to the stable manifold of $\mathbf{x}_F$. If this control law can be defined, when the system representative point reaches the neighborhood of $\mathbf{x}_F$, the value of the parameter p is changed from p_0 to $p_0 + \delta p_1$, with δp_1 defined by the control law. Then, the next iterate of the map will be closer to $\mathbf{x}_F$, and a new parameter value, $p_0 + \delta p_2$, can be chosen which will produce an iterate even closer to the fixed point. Repeating this process again and again the stabilization of $\mathbf{x}_F$ is achieved, and then the unstable periodic orbit is stabilized too.

A quantitative explanation of this method of control will now be given for the case of a two-dimensional surface of section, which is a case frequently found in practice. In this case, the map $\mathbf{x}_{j+1} = \mathbf{F}(\mathbf{x}_j; p)$, which depends on the parameter p, written in vector form reads

$$\begin{bmatrix} x_{j+1} \\ y_{j+1} \end{bmatrix} = \begin{bmatrix} f(x_j, y_j; p) \\ g(x_j, y_j; p) \end{bmatrix}. \tag{5.1}$$

Moreover, it is assumed that the position of a fixed point for the parameter value $p = p_0$, is known,

$$\mathbf{x}_F = \begin{pmatrix} x_F \\ y_F \end{pmatrix}. \tag{5.2}$$

In the neighborhood of the fixed point, the map, $\mathbf{F}(\mathbf{x}_j; p)$, can be approximated by its linearization around $\mathbf{x}_F$ and p_0:

$$\begin{pmatrix} x_{j+1} - x_F \\ y_{j+1} - y_F \end{pmatrix} = \begin{pmatrix} \frac{\partial f}{\partial x} & \frac{\partial f}{\partial y} \\ \frac{\partial g}{\partial x} & \frac{\partial g}{\partial y} \end{pmatrix} \begin{pmatrix} x_j - x_F \\ y_j - y_F \end{pmatrix} + \begin{pmatrix} \frac{\partial f}{\partial p} \\ \frac{\partial g}{\partial p} \end{pmatrix} (p - p_0) \tag{5.3}$$

where the derivatives are computed at $\mathbf{x} = \mathbf{x}_F$ and $p = p_0$. The knowledge of this map is enough to control the dynamics of the system when a state point, $\mathbf{x}_j$, is close enough to the fixed point, $\mathbf{x}_F$, as to the linearized map to be a good approximation of the the real map.

To control chaos by means of changes in the system parameter, a control law which determines the value of p to be used from the actual position of the phase space point, $\mathbf{x}_j$, is needed. This can be taken to be linear

$$p_j = p_0 - [C_x (x_j - x_F) + C_y (y_j - y_F)]. \tag{5.4}$$

The values of the constants C_x and C_y can be determined by substituting this equation in the linearized equation. Then, the condition for the stabilization of $\mathbf{x}_F$ can be formulated by means of the control matrix which can be written as

$$\frac{\partial \mathbf{F}}{\partial \mathbf{x}} - \left(\frac{\partial \mathbf{F}}{\partial p}\right) \cdot \mathbf{C} = \begin{bmatrix} \frac{\partial f}{\partial x} - \left(\frac{\partial f}{\partial p}\right) C_x & \frac{\partial f}{\partial y} - \left(\frac{\partial f}{\partial p}\right) C_y \\ \frac{\partial g}{\partial x} - \left(\frac{\partial g}{\partial p}\right) C_x & \frac{\partial g}{\partial y} - \left(\frac{\partial g}{\partial p}\right) C_y \end{bmatrix}, \tag{5.5}$$

where the vector $\mathbf{C} = (C_x, C_y)$ has been defined. The unstable point $\mathbf{x}_F$ can be stabilized when the modulus of the eigenvalues of the control matrix are smaller then one [Ott et al. (1990); Shinbrot et al. (1993)]. This provides a procedure to construct control laws by means of appropriate choices of the values of C_x and C_y. It is to be noted that in these equations all derivatives are taken at $\mathbf{x} = \mathbf{x}_F$ and $p = p_0$; therefore the control law will be valid only in a close neighborhood of $\mathbf{x}_F$, and for small changes of the control parameter, $\delta p = p - p_0$. To apply Eq. 5.4, an upper bound P for δp has to be chosen, small enough to the linear approximation to be valid. Then, the system is let to evolve freely until a state $\mathbf{x}_j$ is reached which is close enough to $\mathbf{x}_F$ for the condition

$$|C_x (x_j - x_F) + C_y (y_j - y_F)| < P \tag{5.6}$$

to be fulfilled. When this occurs the control law is switched on and the unstable periodic orbit will be stabilized.

If there is explicit knowledge of the map, $\mathbf{x}_{j+1} = \mathbf{F}(\mathbf{x}_j; p)$ (Eq. 5.1), the fixed point, $\mathbf{x}_F$, as well as the Jacobian matrix $\partial\mathbf{F}/\partial\mathbf{x}$ and the vector $\partial\mathbf{F}/\partial p$ at $\mathbf{x} = \mathbf{x}_F$ and $p = p_0$ can be determined straightforwardly. Then the matrix given by Eq. 5.5 can be written and used to choose the constants of the control equation, Eq. 5.4.

However, there is no need of previous knowledge of the form of $\mathbf{F}(\mathbf{x}_j; p)$ to apply the control method, because the eight numbers necessary to determine $\mathbf{x}_F$, $\partial\mathbf{F}/\partial\mathbf{x}$, and $\partial\mathbf{F}/\partial p$ at $\mathbf{x} = \mathbf{x}_F$ and $p = p_0$ can be obtained from a trajectory of the map, $\{\mathbf{x}_1, \mathbf{x}_2, ..., \mathbf{x}_N\}$ with N large enough. The fixed point $\mathbf{x}_F$ can be obtained by means of the observation of recurrent sequences in $\{\mathbf{x}_1, \mathbf{x}_2, ..., \mathbf{x}_N\}$, as described in Chapter 2 when dealing with the techniques used to extract unstable periodic orbits from chaotic attractors. The sequences $\{\mathbf{x}_j, \mathbf{x}_{j+1}\}$, which correspond to motion in the neighborhood of $\mathbf{x}_F$, are then used to obtain the Jacobian matrix, $\partial\mathbf{F}/\partial\mathbf{x}$, by means of a least squares fit of the coefficients in the matrices $\mathbf{A}$ and $\mathbf{B}$ defined by $\mathbf{x}_{j+1} = \mathbf{A} \cdot \mathbf{x}_j + \mathbf{B}$; i.e.

$$\begin{pmatrix} x_{j+1} \\ y_{j+1} \end{pmatrix} = \begin{pmatrix} a_{xx} & a_{xy} \\ a_{yx} & a_{yy} \end{pmatrix} \begin{pmatrix} x_j \\ y_j \end{pmatrix} - \begin{pmatrix} b_x \\ b_y \end{pmatrix}, \tag{5.7}$$

which provides the approximation $\partial\mathbf{F}/\partial\mathbf{x} \approx \mathbf{A}$. The rate of change of the position of the fixed point, $\mathbf{x}_F$, when p changes around p_0 can also be computed approximately by means of the determination of the fixed point coordinates for $p = p_0 + \delta p$, with δp a small change of the system parameter made around p_0. Then,

$$\frac{\partial\mathbf{F}}{\partial p} = \begin{pmatrix} df/dp \\ dg/dp \end{pmatrix} \approx \frac{1}{\delta p} \begin{bmatrix} x_F(p_0 + \delta p) - x_F(p_0 + \delta p) \\ y_F(p_0 + \delta p) - y_F(p_0 + \delta p) \end{bmatrix}, \tag{5.8}$$

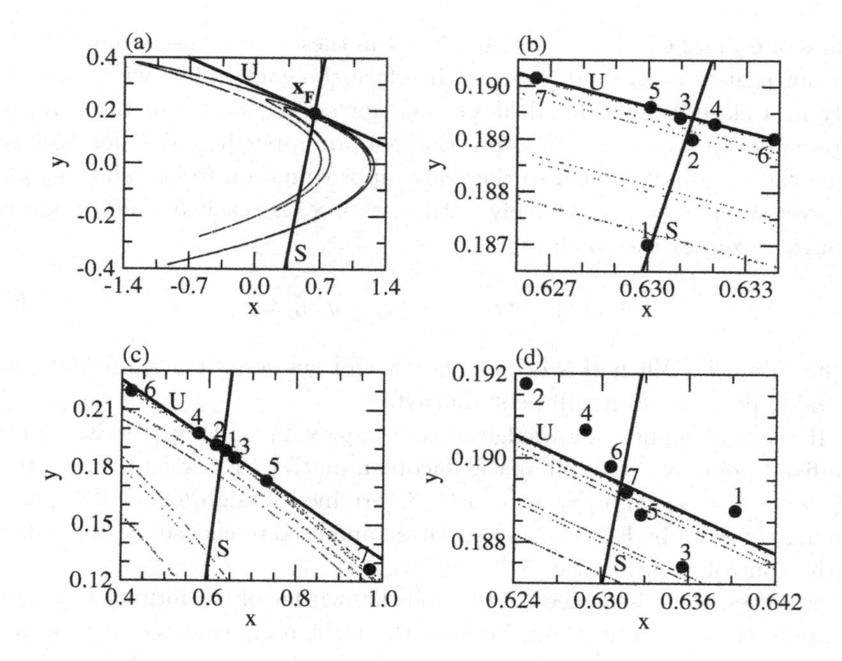

Fig. 5.2 (a) The attractor of the Hénon map (small dots) with the unstable periodic orbit, $\mathbf{x}_F$, represented by a filled circle. Sequences of seven consecutive iterates of the map represented by circles labeled $j = 1, 2, ..., 7$ for: (b) an initial condition very close to $\mathbf{x}_F$ and practically on the stable manifold, (c) an initial condition only close to $\mathbf{x}_F$ and out of any manifold, and (d) the same initial condition as in (c), but now changing the value of the system parameter p at each iterate according to Eq. 5.12. In all these figures the unstable and stable directions are represented by thick lines labeled U and S respectively, which intersect at $\mathbf{x}_F$.

provides this derivative. After these numerical estimations, C_x and C_y can be chosen in the same way as if they had been obtained from the equations of motion.

The control of chaos by means of the OGY method will now be illustrated by means of the example of the control of the Hénon map [Hénon (1976)],

$$x_{j+1} = p - a \cdot x_j^2 + y_j, \tag{5.9}$$

$$y_{j+1} = b \cdot x_j, \tag{5.10}$$

whose dynamics for the parameter values $p = 1$, $a = 1.4$, and $b = 0.3$ occurs in a chaotic attractor. For these parameter values there is an unstable period one orbit, $\mathbf{x}_F$, with coordinates $x_F \approx 0.631354$ and $y_F \approx 0.189406$.

The chaotic dynamics can be controlled by means of the stabilization of this fixed point through small perturbations, δp, applied to the parameter p from the unperturbed value $p_0 = 1$. The Hénon attractor, its unstable periodic orbit, and the correspondent stable and unstable manifolds are displayed Fig. 5.2(a). A sequence of seven consecutive iterates, labeled by its time of occurrence, and starting very close to the stable manifold (point 1) illustrates the dynamics near $\mathbf{x}_F$ in Fig. 5.2(b). First the representative point moves along the stable manifold towards the fixed point (2, 3). Once the unstable manifold is reached, the system diverges exponentially from $\mathbf{x}_F$ along this manifold, alternating positions at the right (4, 6), and at the left (5, 7) of the fixed point. In general [Fig. 5.2(c)], an arbitrary point, not in any manifold, but close to $\mathbf{x}_F$ (point 1) will fall down to the unstable manifold (2, 3), move away from the fixed point along the unstable manifold (4, 5), and when far enough will leave the manifold (6, 7). This is how the instability manifests itself. This behavior, however, can be controlled by changing the values of the system parameter p at each iteration in an appropriate way. This is illustrated in Fig. 5.2(d) where a sequence of seven points is displayed, started at the same initial condition as that in Fig. 5.2(c), and with the parameter p changing at each iteration along the series of values: 1.013, 1.008, 0.990, 1.007, 0.996, 1.002, 0.999. Now, the sequence of phase space points moves towards the fixed point, which means that this is stabilized.

The procedure to achieve this stabilization requires the definition of a control law (Eq. 5.4) by means of the analysis of the control matrix (Eq. 5.5). For the Hénon map the control matrix to be analyzed is

$$\frac{\partial \mathbf{F}}{\partial \mathbf{x}} - \left(\frac{\partial \mathbf{F}}{\partial p}\right) \cdot \mathbf{C} = \begin{pmatrix} -2ax_F - C_x & 1 - C_y \\ b & 0 \end{pmatrix}, \tag{5.11}$$

whose eigenvalues can be made equal to $\Lambda_1 = \Lambda_2 = -ax_F/2$ just by means of the choice $C_x = -ax_F$, and $C_y = 1 + (ax_F)^2/4b$. For the particular system parameter values considered here, the numerical values of the eigenvalues are $\Lambda_1 = \Lambda_2 \approx -0.441\,95$ which have an absolute value smaller than one; therefore,

$$p_j = 1 + ax_F(x_j - x_F) - \left[1 + \frac{(ax_F)^2}{4b}\right](y_j - y_F), \tag{5.12}$$

defines an appropriate control law for the Hénon map in the present example.

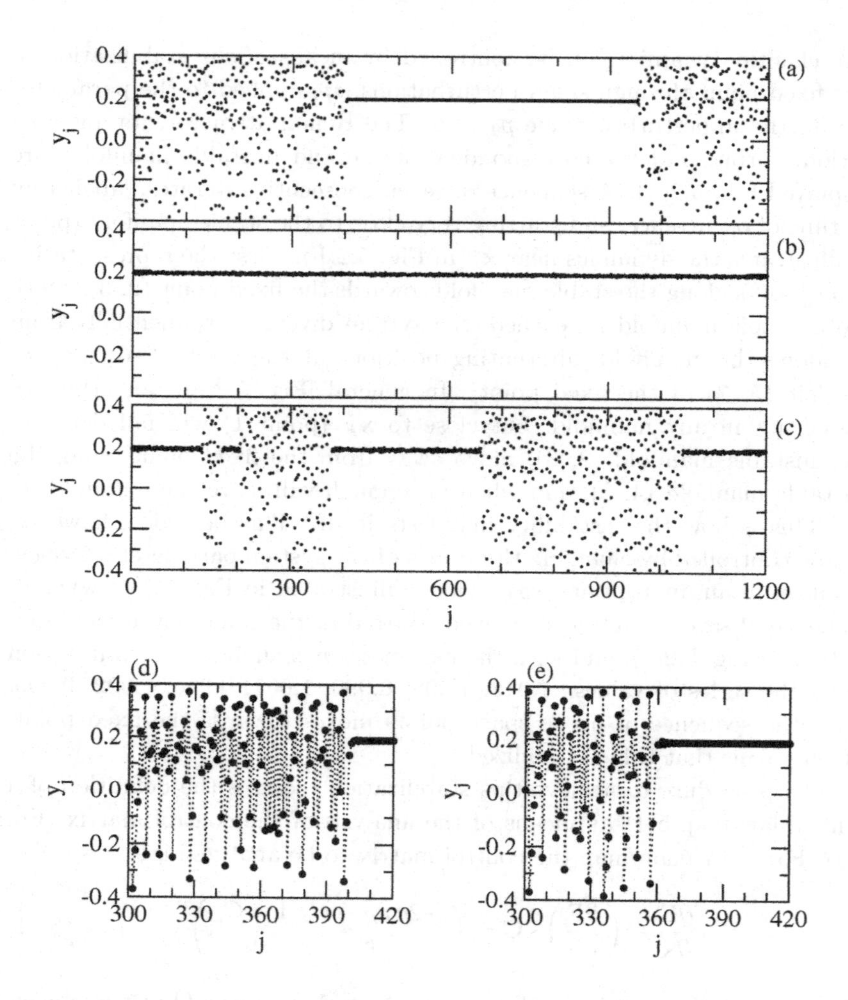

Fig. 5.3 (a) Control of chaos in the Hénon map monitored by means of the variable y. Effect of a Gaussian noise in the control of chaos for different levels of noise: (b) $\sigma = 0.005$, and (c) $\sigma = 0.060$. Transitory evolutions since the control device is switched on, until chaos is controlled for (d) $P = 0.04$ and (e) $P = 0.08$.

Control of chaos in the Hénon map is further illustrated in Fig. 5.3 by means of time series of the variable y. An example of stabilization phenomenon is presented in Fig. 5.3(a) by means of a time series of 1200 iterates of the map from an arbitrary initial condition chosen in the stable attractor, and setting $P = 0.04$. The system was left to evolve freely, showing a typical chaotic dynamics, until the iteration $j = 300$ when the control procedure was initiated. During an initial transitory evolution of nearly 100 iterates the system remains chaotic, until the chaotic orbit passes close enough to $\mathbf{x}_F$ for the condition in Eq. 5.6 to be fulfilled. Chaos is then controlled by means of the control law given by Eq. 5.12, and y_j remains approximately equal to y_F. At $j = 900$, control is switched off, and after approximately 50 iterations, the instability becomes dominant, and chaos develops again.

The control of chaos achieved in this way is robust enough to be observed in experiments. This is illustrated in Figs. 5.3(b, c) where the effect of external noise on the previously controlled system, subject to a Gaussian noise is presented. The noise, which acts as an additive term in each of Eqs. 5.9–5.10, has a distribution with zero mean, and its strength is measured by the standard deviation of the distribution. As shown in Fig. 5.3(b) a small amount of noise does not alter the controlled state, and only causes a small fluctuation of the system variables. When the strength of the noise increases [Fig. 5.3(c)], sporadic burst where control is lost develop. As noise strength increases further the bursts increase in frequency and length, and for a noise level large enough there is a complete loss of control.

As noted above, the control of chaos is not achieved immediately after switching on the control mechanism; rather, as shown in detail in Fig. 5.3(d), there is a transient before the system dynamics becomes effectively controlled. It has been proven [Ott et al. (1990)] that the relation between P and the average length of a transitory, T are related by a potential law, $T = P^{-\gamma}$. That transitory can then be shortened by increasing the value of P, which makes the condition Eq. 5.6 less restrictive [Fig. 5.3(e)]. There is, however, a restriction to this possibility because P has to remain small enough for the linear approximations made to be valid. When there is a need to drastically reduce this transitory, a technique devised to direct trajectories to a prescribed accessible state in a short time [Shinbrot et al. (1990); Shinbrot et al. (1993)] can be used. This method, which relies on the sensitivity to initial conditions proper of chaotic systems, also proceeds by the application of small appropriate perturbations to a single system parameter.

Finally, it is worth noting, that in the example of the Hénon map presented here, the control law (Eq. 5.12) was determined in a heuristic way by means of the examination of the expression that gives the eigenvalues of the control matrix (Eq. 5.11). A systematic procedure can also be used [Ott et al. (1990)] that requires the knowledge of the unitary vectors in the direction of the stable and unstable manifolds and the numerical values of the instabilities for each manifold. All this can be obtained in a straightforward manner when the system equations are known, and can also be determined experimentally from the observation of system trajectories like those presented in Figs. 5.2(b, c).

5.1.2 *Experiments and applications*

The OGY method was initially [Ott et al. (1990); Shinbrot et al. (1993)] introduced and illustrated by means of computer experiments performed on the Hénon attractor, for parameters different to those used here, and with somewhat different choices for the control law. After its introduction the OGY method, and certain modifications, have proven to be useful to control chaos in the real world.

Experimental control of chaos using the OGY method has been achieved in a magneto-elastic system [Ditto et al. (1990)]. This was made of a vertical elastic ribbon clamped at its base, and bent by its weight [Fig. 5.4(a)]. The elastic properties of the material the ribbon was made of were dependent on the magnetic field applied along a nonlinear law. Then, the strain of the ribbon could be modified by applying an external magnetic field. This simple magneto-mechanical system can behave like a chaotic oscillator [Savage et al. (1990)] when it is inserted in a system of coils which compensate for the Earth's magnetic field, and create a new controllable vertical field whose intensity changes sinusoidally with time: $H(t) = H_0 + H_1 \cos(\omega t)$, with H_0, H_1 and ω constant values. The output of the experiment is a measure of the position of a point of the beam as a function of time. The chaotic properties of this oscillator can be studied by means of standard embedding techniques; and, in particular Poincaré maps, $\mathbf{x}_{j+1} = \mathbf{F}(\mathbf{x}_j)$, can be reconstructed. The experimental control of chaos in this system was experimentally achieved [Ditto et al. (1990)] by parametric perturbations of no more than nine percent of the constant vertical field H_0. The control law was designed along standard prescriptions found in the literature [Ott et al. (1990)], and its application to the chaotic elastic ribbon allowed to alternate the stabilization of a period one, and a period two orbits at will.

A variation of the OGY method based in the use of return maps, instead of Poincaré maps, has been used to control chaos in chemical reactions [Peng et al. (1991); Petrov et al. (1993)]. In this case it was assumed that a one-dimensional return map, $x_{j+1} = F(x_j; p)$, dependent on a system parameter p, can be extracted from the observed dynam-

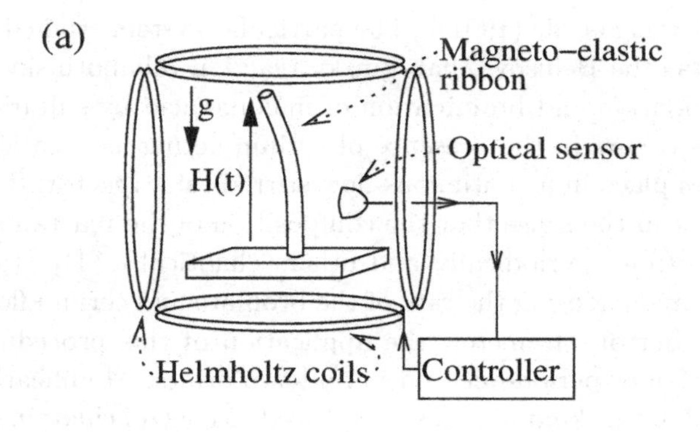

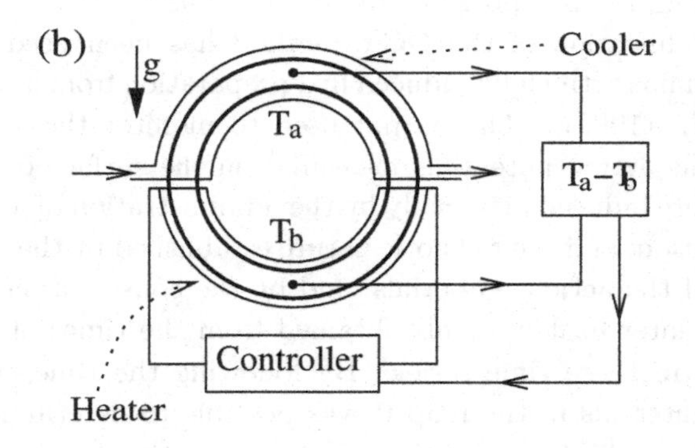

Fig. 5.4 Schemes of two experiments where control of chaos has been achieved by means of the OGY method. (a) A magneto-elastic ribbon is left free of the Earth's magnetic field by means of a set of Helmholtz coils which create a compensating magnetic field plus an additional perpendicular field, $\mathbf{H}(t)$. The deviation of the beam from the vertical line is measured by an optical sensor which sends a signal to a controller which modifies the strength of $\mathbf{H}(t)$. (b) A torus shaped vessel (thick line), containing a liquid, in the gravitational field is cooled from above and heated from below. The values of the temperature difference between two points in the vessel are used to generate a signal which is used to control the rate of heating.

ics. This is linearized around the fixed point of interest, x_F, to obtain $x_{j+1}(p) = a \cdot [x_j - x_F(p)] + x_F(p)$, where $a = \partial F/\partial x$, taken at x_F and p, is a scalar quantity. The derivative $b = \partial F/\partial p$, at x_F and p, is another scalar which provided the change of the position of the fixed point with p. The control law to apply when the system state is, $x_j = x_F + \delta x$, close to x_F is then $\delta p = C \cdot \delta x$, with $C = a/[(a-1)b]$, so that the perturbation to be applied to p is just proportional to the distance to the fixed point to be stabilized [Peng et al. (1991)]. The particular system studied [Petrov et al. (1993)] was the Belusov–Zhabotinsky reaction [Zhabotinsky (1991)] in which the oxidation and bromination of malonic acid by sulfuric acid and bromate ions occurs in the presence of cerium acting as a catalyzer. The reaction takes place in a continuous-flow stirred-tank reactor. This system is an oscillator in the sense that the composition of the reactants oscillates in time, sometimes periodically and others chaotically. By starting in a chaotic state and choosing the rate of the bromate and cerium flow into the tank as the control parameter, the application of this procedure allowed the stabilization of period-one and period-two orbits. Modifications of the OGY method of this kind have also been used to control chaos in electronics [Hunt (1991)], and in laser physics [Roy et al. (1992)].

Another modification of the OGY method has been used to control cardiac arrhythmias artificially induced in a preparation from a rabbit heart [Garfinkel et al. (1992)]. The output used to monitor the system state were monophasic action potentials measured on the surface of the tissue. Arrhythmias were induced artificially by the administration of a drug to the cardiac preparation, and their chaotic nature established by the observation of time series of the action potentials, and by the construction of discrete maps from the inter-beat intervals obtained from the times of occurrence of the maxima in these time series. By following the time evolution of the inter-beat intervals in the map it was possible to determine the fixed points, and their stability properties to be used in the design of a control mechanism. Because in this case there were no system parameter available, short constant voltage pulses applied to the heart were the input used to control chaos. The control strategy designed, which is a modification of the OGY method known as proportional perturbation feedback, is aimed to determine the time for which an electric pulse of fixed size should be applied. A control theory aimed to bring the system state point to the stable manifold, was developed [Garfinkel et al. (1992)], whose mathematical formulation resulted in control equations identical to those of the OGY method [Ott et al. (1990)], with the time interval between beats used as the control parameter. By means of this method it was possible to convert regimes of cardiac arrhythmia into periodic regimes given by period three orbits.

This modification of the OGY method has also been applied to control the bursting behavior of neuronal populations [Schiff et al. (1994)]. The systems studies in this case were in vitro hippocampal slices, whose electrical activity were measured to determine inter-burst intervals. These were used to establish the nature of the bursting, to localize the unstable orbits, and to fit their instabilities as in the above case of the cardiac tissue. The application of proportional perturbation feedback, using single or double pulses, worked to diminish the degree of chaos, and even to suppress chaos in several experiments.

Suppression of turbulence in fluid convection by means of small perturbations of a system parameter has also been reported [Singer et al. (1991)]. The experimental system was an amount of water, filling a vessel which had the shape of a torus [Fig. 5.4(b)]. This was disposed with the plane containing the main circle of the torus placed vertically. The upper half of the torus was cooled by an external flow of water, and the lower half uniformly heated by an electrical resistance. This produced a convective circulation of the fluid along the torus, which for a heating rate large enough happened to be turbulent. The heating rate, W, was chosen as the control parameter, and fixed changes, δW, of less then five percent of its nominal value were applied to laminarize the fluid. Instead of constructing a Poincaré map to design a control law like Eq. 5.4, these authors [Singer et al. (1991)] measured the temperature difference between the top and bottom of the vessel, and applied the perturbation δW when this difference was over its average value, and $-\delta W$ when it was below. This control procedure was enough to turn the turbulent fluid dynamics to a laminar one, i.e. to control chaos. Although, in this case, the control scheme was not designed from an analysis of fixed points and their stabilities, the method presents certain resemblance with the proportional perturbation feedback, in the sense that the perturbation is designed to move the system state towards the fixed point, instead of the stable manifold towards the state point, as in the original OGY method.

5.2 The Pyragas method

A method has been developed [Pyragas (1992)] to stabilize unstable periodic orbits applying small time continuous control to a parameter of a system while it evolves in continuous time, instead of a discrete control at the crossing of a surface. This is known as delayed feedback control, and numerical simulations and experiments have proven it to be easy to implement, and effective at least for the less unstable periodic orbits; i.e. the orbits with smaller periods.

5.2.1 *Delayed feedback control*

The Pyragas method assumes that the system evolves in a continuous time. This is the kind of system described by variables $\mathbf{x} = (x_1, x_2, ..., x_d) \in \mathbb{R}^d$ whose time evolution is given by an autonomous nonlinear flow

$$\frac{d\mathbf{x}}{dt} = \mathbf{F}(\mathbf{x};p), \tag{5.13}$$

with p an externally controllable parameter whose numerical value is zero in the absence of external perturbations. It is assumed that for $p = 0$ the attractor is in the chaotic state of interest, whose periodic orbits are to be stabilized. Moreover, it is assumed that there is an observable of the system, which is a scalar signal, $s(t)$, give by some function of the state of the system, $s(t) = \mu[\mathbf{x}(t)]$, when a measure of that state is performed in experiments.

An unstable periodic orbit, of period τ, of this system which verifies $\mathbf{x}_p(t+\tau) = \mathbf{x}_p(t)$, can be stabilized by means of the delayed feedback control, with no additional information on this orbit, by modifying the control parameter p along the following control law

$$p(t) = C \cdot [s(t-\tau) - s(t)], \tag{5.14}$$

where C measures the strength of the perturbation, and $s(t-\tau)$ is the signal measured with a time delay equal to τ. If τ is the period of an unstable periodic orbit embedded in the attractor, by means of a proper choice of the value of C, an unstable periodic orbit with period τ may be stabilized. When stabilization is achieved, according to Eq. 5.14, the control parameter recovers its nominal value, $p(t) = 0$. Although the dynamical behavior that corresponds to this parameter value is the chaotic attractor whose control was aimed, the trajectory effectively followed is a periodic orbit of period τ which has to be a solution of Eq. 5.13 for $p = 0$. This means that the control law has not created a new orbit; rather, it just has stabilized an existing unstable one. Therefore, chaos has been controlled by means of the stabilization of one of the unstable periodic orbits of the system.

Once control has been achieved, the size of the perturbation, $p(t)$, needed to maintain it is very small; however, during the previous transitory this can be quite large. If the size of the perturbation is limited by some value, P, the control of chaos can still be achieved by means of the bounded control law

$$p(t) = \begin{cases} -P & , \text{if } C \cdot [s(t-\tau) - s(t)] \leq -P, \\ C \cdot [s(t-\tau) - s(t)] & , \text{if } |C \cdot [s(t-\tau) - s(t)]| < P, \\ P & , \text{if } C \cdot [s(t-\tau) - s(t)] \geq P, \end{cases} \tag{5.15}$$

which, in the neighborhood of the periodic orbit, reduces to Eq. 5.14. This control scheme is advantageous in systems that have multiple basins of attraction because it reduces, and even avoids, the possibility of the stabilization of an orbit different than the one of interest [Pyragas (1992)]. In turn, the bounded control law has the disadvantage of making transitories longer as P is decreased.

The Duffing oscillator [Moon (1987)] equation,

$$\frac{d^2x}{dt^2} + \gamma\frac{dx}{dt} - \frac{1}{2}x\left(1 - x^2\right) = A\sin(t), \tag{5.16}$$

which describes the motion of a damped oscillator in a double-well potential under the action of a harmonic force, will now be used to present an example of control of chaos by delayed feedback control. For the parameter values $\gamma = 0.10$ and $A = 0.24$ the systems dynamics occurs in a chaotic attractor, whose projection onto the two-dimensional phase space defined by the position and the velocity, x and $y = dx/dt$, is presented in Fig. 5.5(a). To study the control of chaos, the oscillator equation is rewritten under the form of a continuous flow

$$dx/dt = y, \tag{5.17}$$

$$dy/dt = -\gamma y + \frac{1}{2}\left(x - x^3\right) + A\sin(z) + C\left[y(t-\tau) - y(t)\right], \tag{5.18}$$

$$dz/dt = 1, \tag{5.19}$$

where the control law, defined by the parameters τ and C, is introduced as a second additional force with feedback. For this system, the less unstable periodic orbit is a period one orbit which has the frequency of the periodic force. Therefore, it may be stabilized by setting $\tau = 2\pi$, and choosing an appropriate value for the stiffness of the controlling force, C.

In particular, for $C = 0.36$ the system trajectory obtained after transitories have died off is presented in Fig. 5.5(b), and shows a simple period one cycle instead of the complex entangled trajectory seen in Fig. 5.5(a), despite the system parameters being the same. This periodic orbit has been obtained from a run which spans a time period equal of 500 [Fig. 5.5(c)]: until $t = 50$, the system was left to evolve freely (i.e. $C = 0$), then the

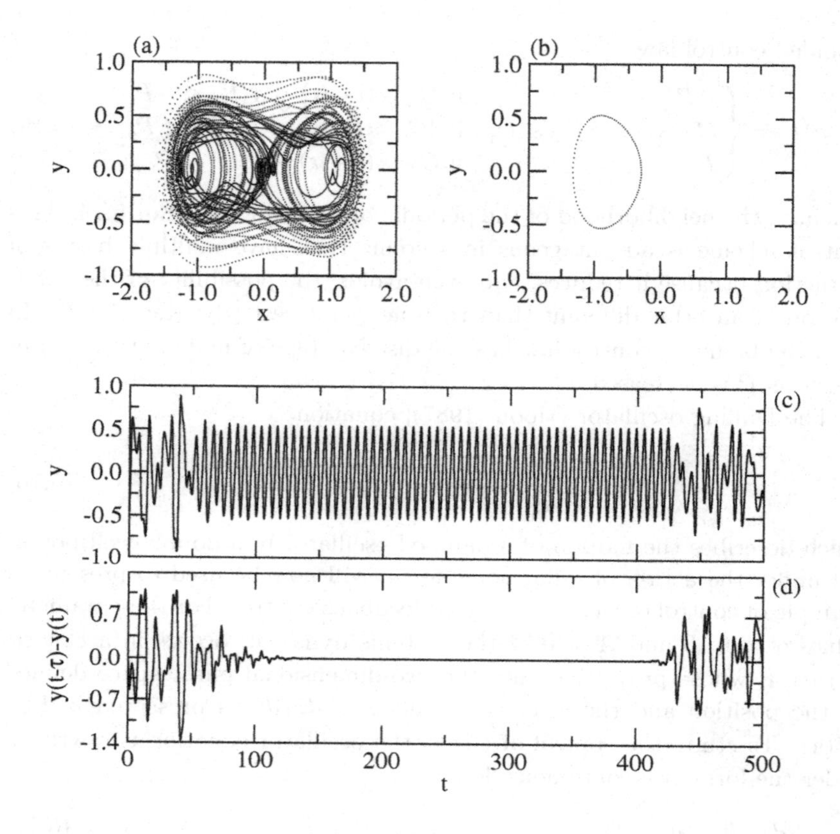

Fig. 5.5 Trajectories for a Duffing oscillator ($\gamma = 0.10$, $A = 0.24$) in the position-velocity plane: (a) evolving freely in the chaotic regime, and (b) subject to a continuous delayed feedback control tuned at $\tau = 2\pi$, and $C = 0.36$ (after transitories have died off). An event of chaos control monitored by: (c) a time series of $y = dx/dt$, and (d) a time series of the difference $y(t-\tau) - y(t)$; control is switched on at $t = 50$, and switched off at $t = 400$.

control was switched on ($C = 0.36$), until $t = 400$ when it was disconnected again. The effect of the control is indeed to stabilize a periodic orbit for $t = 50$, which is unstable as it loses its stability at $t = 400$ when the control is removed. The correspondent time evolution of the quantity $y(t-\tau) - y(t)$, presented in Fig. 5.5(d), shows that the controlling force practically vanishes, and stays null, once the orbit has been stabilized. This means that when the control is working the time evolution occurs according to Eq. 5.16, so that the periodic orbit shown in Fig. 5.5(b) is a solution of the same equation whose stable solution is the chaotic trajectory shown in

Fig. 5.5(a); i.e. it is one of the unstable periodic orbits embedded in the chaotic attractor.

To obtain a quantitative characterization of the quality of the control of chaos achieved, the following integral can be computed:

$$D(\tau, C) = \lim_{T \to \infty} \frac{1}{T} \int_0^T |y(t-\tau) - y(t)|\, dt. \tag{5.20}$$

This measures the time average of the absolute value of the difference, $y(t-\tau) - y(t)$, for an event of chaos control. When an unstable periodic orbit is controlled, the control force, Eq. 5.14, is practically zero; therefore, $D(\tau, C)$ has to be null. If $D(\tau, C)$ is not null the control law has to be permanently active, the dynamical equation (Eq. 5.13) does not reduce to that of the unperturbed oscillator, and no warranty of stabilization of an unstable periodic orbit of the system exists.

Because $D(\tau, C)$ depends on the period of the orbit to be stabilized, τ, and on the stiffness of the perturbation, C, a study of these dependencies is relevant. The dependence on τ is presented in Fig. 5.6(a), for three values of C. This shows that, for the weakest couplings, this quantity stays well above zero for all values of τ indicating that no unstable periodic orbit is being controlled. For C large enough $D(\tau, C)$ presents an acute minimum at the period of the unstable periodic orbit, $\tau = 2\pi$, when $D(\tau = 2\pi, C) = 0$; i.e. only in this case there is stabilization of one of the unstable periodic orbits of the Duffing oscillator. The dependence on $D(\tau, C)$ for $\tau = 2\pi$ [Fig. 5.6(b)] shows that here is a threshold for the development of the control of chaos which in this example occurs for $C \approx 0.30$. For values of τ in the vicinity of 2π the condition $D(\tau, C) = 0$ is never fulfilled, this indicating that the attractors obtained are not solutions of the unperturbed system.

Finally, it is to be noted that the control achieved by this method is robust against external noise. This is illustrated in Figs. 5.6(c, d), which display an event of control of chaos like that on Figs. 5.5(c, d), but in this case in a noisy environment which has been simulated by the addition of a Gaussian noise of variance σ to Eq. 5.18. For the case plotted in Figs. 5.6(c, d), $\sigma = 0.04$, which is about ten per cent of the amplitude of the oscillation of dy/dt, measured as a function of time in the chaotic regime, as given by Eq. 5.18. The effect of this noise is mild, its main effects being to make the transitory large since the control is connected at $t = 50$, and to accelerate the departure form the periodic orbit when it is disconnected at $t = 400$. A neat periodic orbit is obtained when the control has been achieved in the interval $150 < t < 400$.

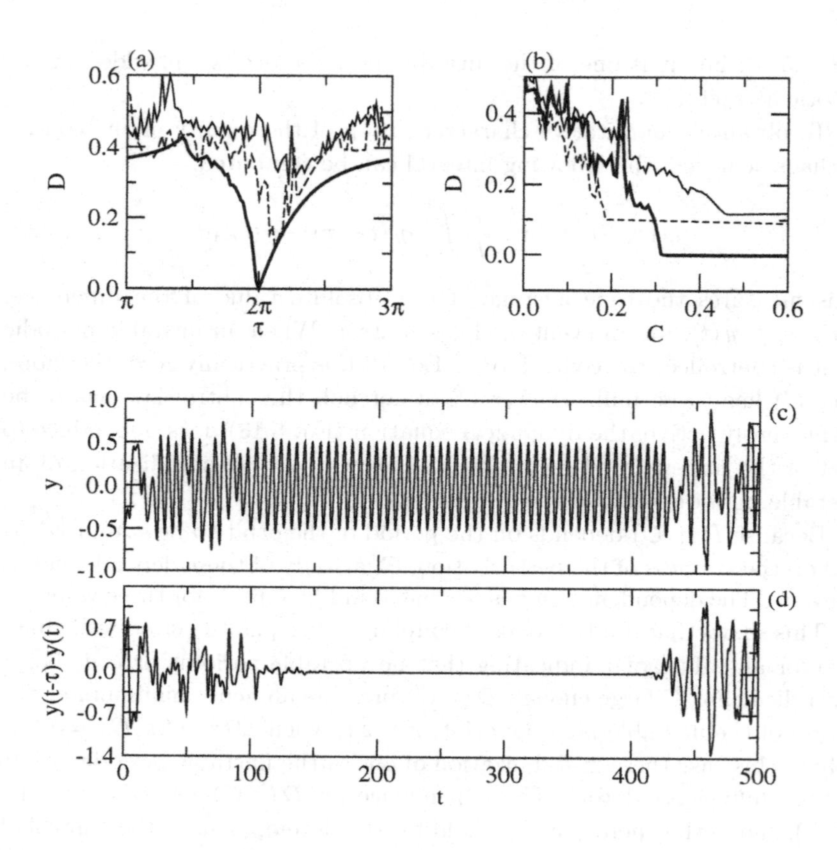

Fig. 5.6 The time average of the difference $|y(t-\tau) - y(t)|$ as a function of the parameters of the control law: (a) dependence on τ for $C = 0.36$ (thick line), $C = 0.18$ (dashed line) and $C = 0.06$ (thin line), and (b) dependence on C for $\tau = 2\pi$ (thick line), $\tau = 0.95 \cdot 2\pi$ (dashed line), and $\tau = 1.05 \cdot 2\pi$ (thin line). An event of chaos control in the presence of a Gaussian noise with standard deviation $\sigma = 0.040$ monitored by: (c) a time series of $y(t)$, and (d) a time series of $y(t-\tau) - y(t)$ with control switched on at $t = 50$, and switched off at $t = 400$.

5.2.2 *Experimental realizations*

The first events of control of chaos by means of delayed feedback control have been numerical experiments [Pyragas (1992)] performed mainly on the Rössler model. However several experimental realizations have followed which show the usefulness of this method in the laboratory.

The control of chaos by means of delayed feedback control has been accomplished in a magneto-mechanical system [Hikihara and Kawagoshi

(1996)]. This was made of a vertical steel beam, fixed at its upper tip, and hanging over the middle point of the line joining two permanent magnets close to the lower tip [Fig. 5.7(a)]. The whole system was periodically shaken, so that a good mathematical model for it was that of a damped oscillator, subject to a potential determined by the position of the magnets, and acted by an harmonic force, similar to that described by Eq. 5.16. The potential created by the magnets could be made to be a two well potential (like that of Eq. 5.16), or a three well potential by changing the relative distances between magnets and beam. By measuring the position of the beam by means of an optical sensor, it was possible to observe the state of the oscillator, and then to determine the values of the forcing frequency and amplitude where chaotic vibrations develop. Working in the chaotic regime, it was possible [Hikihara and Kawagoshi (1996)] to stabilize orbits of period one and of period two in the three well potential, and of period one in the two well potential. To achieve this control a straightforward application of the delayed feedback control was made: the perturbation given by Eq. 5.14 was added to the oscillator controlling the shaker. Moreover, the inverse of the forcing frequency was used as the delay time for period one orbits, and twice this value for period two orbits.

Stabilization of unstable periodic orbits in a chemical system has also been reported [Parmananda et al. (1999)]. The particular system studied was a standard electrochemical cell with a copper anode, a platinum cathode, and electrolyte made of a mixture of acetic acid and sodium acetate. The anodic current was the scalar signal used to monitor the state of the system, and to assemble the delayed feedback control (Eq. 5.14) to be used to modulate the anodic potential, which was the control parameter chosen. With this experimental setup these authors [Parmananda et al. (1999)] reported the stabilization of periodic orbits and fixed points from unperturbed chaotic states, as well as of fixed points from unperturbed periodic states by means of the variation of the time delay in the control law. Moreover, these observations were substantiated by computer simulations of a model for electrochemical corrosion [McCoy et al. (1993)] which were found in qualitative agreement with the experiments.

Chaos in fluid mechanics has also been controlled by means of delayed feedback control [Lüthje et al. (2001)]. The experimental system was a Taylor–Couette flow which is the motion of a fluid in the space between two vertical concentric cylinders, closed at their top and bottom by flat plates. The inner cylinder, which rotates with a certain angular velocity,

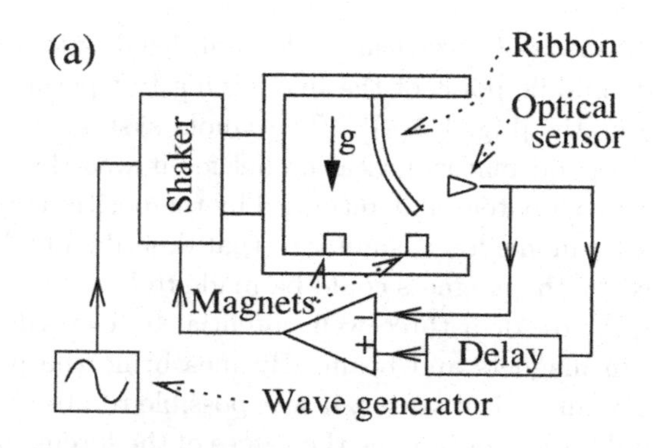

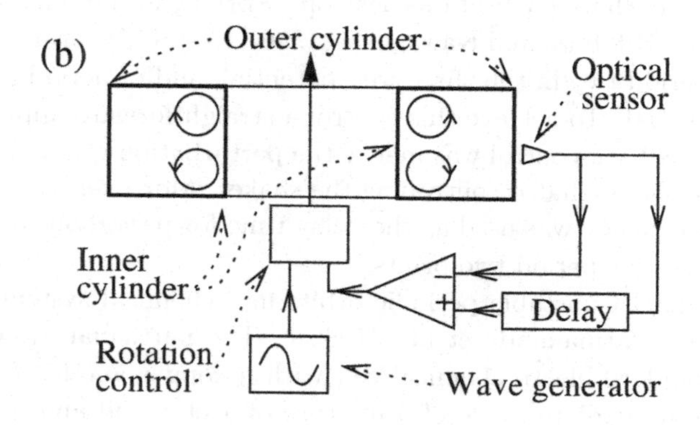

Fig. 5.7 Experimental realization of delayed feedback control. (a) A metallic ribbon hangs vertically between two magnets. This system is shaken, and an optical sensor measures the deviation of the ribbon from the vertical. This output is used to modify the shaking rate. (b) A fluid is enclosed between two vertical concentric cylinders and two horizontal plates (thick lines). The outer cylinder is fixed while the inner one rotates around its axis. An optical sensor measures the vertical velocity of the fluid and this output is used to modify the rotation rate of the inner cylinder.

causes the motion of the viscous fluid between the cylinders. The angular velocity is the system control parameter. For appropriate values of the angular velocity, a fluid dynamics regime develops in which horizontal vortex develop. This fluid dynamics regime is known as the Taylor vortex flow [Fig. 5.7(b)]. The vertical velocity of a vortex measured by a Doppler optical sensor is the scalar variable used to monitor the state of the system, which for large enough values of the angular frequency becomes chaotic. This signal is then used to prepare a bounded control law, like Eq. 5.15, to be acted on the mechanism controlling the angular velocity of the inner cylinder. The delay was chosen as the inverse of the frequency of the main peaks of the power spectral density observed in the chaotic regime. For appropriate values of the stiffness of the perturbation it was found that period one and period two orbits could be stabilized [Lüthje et al. (2001)]. A study of the dependence of the control of chaos on the perturbation stiffness showed that control developed only in a finite bounded interval, $C_{\min} < C < C_{Max}$.

A modification of the delayed feedback control has been proposed [Pyragas (1992)] which, given the system modelled by Eq. 5.13, assumes that the periodic orbit to be stabilized, $\mathbf{x}_P(t)$, has been extracted from the attractor through the standard techniques invented to this aim, that have been presented in Chapter 2. The scalar signal $s_P(t) = \mu[\mathbf{x}_P(t)]$ can then be obtained, and the unstable periodic orbit is stabilized by applying the control law $p(t) = C \cdot [s_P(t) - s(t)]$ to the control parameter p with an appropriate value of stiffness of the perturbation, C. As before, once the orbit has been stabilized, the perturbation becomes null.

This technique has been tested successfully by means of numerical simulations of the Rössler and Lorenz models [Pyragas (1992)]. However, the implementation of this approach in experiments is harder than delayed feedback control. This is because it requires two additional tasks. For one side the laborious work of extracting the unstable periodic orbit, $\mathbf{x}_P(t)$, of the system has to be performed first. Then, there is the need to design an oscillator which creates the signal $s_P(t)$ to be used in the control law. The delayed feedback control approach, that only requires the knowledge of the orbit period, is then simpler and has been preferred by the experimentalist willing to control chaos by continuous control [Pyragas (2002)]. For this reason, delayed feedback control has also received more attention from the theoreticians who have proposed several improvements to increase its stabilization power, i.e. the range and nature of periodic orbits that can be stabilized [Socolar et al. (1994); de Sousa Vieira and Lichtenberg (1966)].

Chapter 6

Mutually Coupled Identical Chaotic Oscillator

Systems of chaotic oscillators mutually coupled are frequently found in the laboratory and in the natural world. In many cases, two oscillators are enough to model the system of interest; while in others, networks of several or many oscillators are found. In any case, for coupling strength large enough it is possible to observe dynamical states which are coherent in the sense that all oscillator evolve in synchrony. Moreover, for lesser coupling a complex phenomenology may develop. The case when these oscillators are identical or nearly identical will be studied in this chapter, the case when the oscillators have different individual dynamics will be considered in the next chapter. In this chapter, first, the description of the synchronized state achieved, and a study of its stability will be presented for a set of two mutually coupled identical oscillators. Then the possible complexities developed in the transition to the synchronized state will be discussed. This will be followed by an account of numerical and experimental observations made by several authors. The manifestation of these phenomena in networks of identical chaotic oscillators will, finally, be outlined together with a discussion of the particular phenomenology that may develop in such networks, known as partial synchronization of chaos.

6.1 Synchronization of two coupled oscillators

Very frequently in science and technology, complex systems are composed of simpler units. Complicated electric circuits that are synthesized by the mutual connection of two or more simpler circuits, or systems of several interacting lasers are examples that are found in physics and engineering. In neurobiology single neurons and groups of neurons that interact through dendritic connections provide an example of coupled chaotic oscillators of interest in biology. Also in chemistry and fluid dynamics, networks of coupled oscillators are interesting because they can be used to model spatially extended systems in terms of the local behavior around properly chosen nodes distributed along the system's spatial extension. The simplest possible network is formed by two identical coupled oscillators, and will be studied in this section. Besides being a minimal model useful to analyze and understand the collective behavior of larger and more complex systems, it is interesting in itself because many cases found in practice refer to only two coupled oscillators.

6.1.1 *Synchronized motion and its stability*

Research on the synchronization of identical chaotic oscillators has focused mainly in autonomous chaotic systems, with a dynamics governed by a vector field, $\mathbf{F}(\mathbf{x})$, in a phase space of dimension d; i.e. $\mathbf{x} \in \mathbb{R}^d$. When two identical, labeled 1 and 2, mutually coupled chaotic oscillators have been studied, a case that has received major attention is [Fujisaka and Yamada (1983)]:

$$\frac{d\mathbf{x}_1}{dt} = \mathbf{F}(\mathbf{x}_1) + \mathbf{C} \cdot (\mathbf{x}_2 - \mathbf{x}_1), \tag{6.1}$$

$$\frac{d\mathbf{x}_2}{dt} = \mathbf{F}(\mathbf{x}_2) + \mathbf{C} \cdot (\mathbf{x}_1 - \mathbf{x}_2), \tag{6.2}$$

with $\mathbf{C}$ a symmetric matrix of constants which describes the nature, and strength of the coupling between the oscillators. This is called the interaction matrix. The type of coupling defined by Eqs. 6.1–6.2 is called diffusive coupling in the literature.

An interesting type of motion available to this system is when, given $\mathbf{x}(t)$ which verifies the differential equation $d\mathbf{x}/\mathbf{dt} = \mathbf{F}(\mathbf{x})$ for all t, the trajectories of each oscillator verify

$$\mathbf{x}_1(t) = \mathbf{x}_2(t) = \mathbf{x}(t). \tag{6.3}$$

A state like this is called the macro-oscillation, or also the synchronized state [Fujisaka and Yamada (1983)]. In general, the motion of the system given by Eqs. 6.1–6.2, occurs in a phase space of dimension $2d$; however, if the synchronized state is achieved, the motion collapses to a hyperplane of this phase space defined by the condition $\mathbf{x}_1 = \mathbf{x}_2$, which is called the synchronization manifold. Because it is assumed that the attractor to which $\mathbf{x}(t)$ belongs exists, and $\mathbf{x}_1 = \mathbf{x}_2$ is a geometrical constraint, the synchronized state and the synchronization manifold do always exist. The phase space can then be seen as composed of two geometrical entities: the synchronization manifold, and the subspace orthogonal to it, which is named the transverse subspace. These two geometrical concepts have proven to be useful in the study of the synchronization of mutually coupled chaotic systems [Pecora et al. (1997)].

The relevant problem here is the stability of the synchronization manifold; i.e. the question of what happens when a small perturbation brings a system, which is evolving in the synchronization manifold, out to a state having a non null projection onto the transverse subspace. If the perturbation dies off exponentially, and the perturbed system returns to the synchronization manifold, the synchronized state is said to be stable. If not, it is unstable when the perturbation growths exponentially, or marginally stable when the perturbed trajectory stays wandering in the neighborhood of the synchronization manifold. The existence of stable synchronized states is the main result obtained in the study of the dynamics of two coupled chaotic identical oscillators [Fujisaka and Yamada (1983)]. The description of this synchronized state is essentially the same as what was called identical synchronization in Chapter 4 and can be detected and analyzed in the same way described there.

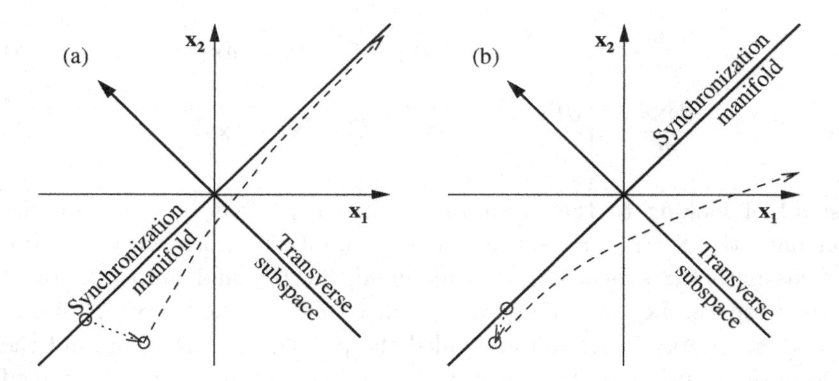

Fig. 6.1 The two phase spaces of the individual oscillators x_1 and x_2 are represented as thin line axes. The synchronization manifold is represented by the thick diagonal line, and the transverse subspace by the thick diagonal axis. Perturbations of a motion in the synchronization manifold are represented by a dotted line that connects the states before and after the perturbation (small circles), and the resulting phase space trajectories are presented as dashed lines. There are two main cases: (a) asymptotic stable manifolds, and (b) asymptotic unstable manifolds.

A schematic representation of the elements implied in the theory of stability of synchronized motion [Fujisaka and Yamada (1983); Pecora et al. (1997)] is presented in Fig. 6.1. The whole phase space can be described as the composition of the two individual phase spaces of the oscillators $\mathbf{x}_1$ and $\mathbf{x}_2$, of dimension d each. Then this has dimension $2d$, and the dynamical state of the coupled oscillators is given by the coordinates $(\mathbf{x}_1, \mathbf{x}_2)$. The synchronization manifold is the hyperplane, of dimension d, that results from the condition $\mathbf{x}_1 = \mathbf{x}_2$, and the transverse subspace is the hyperplane, of dimension d, perpendicular to it. When an infinitessimal perturbation of a motion that occurs in the synchronization manifold has a component in the direction of the transverse subspace this may result in one of two different kinds of trajectories in the whole phase space: if the manifold is asymptotically stable the trajectory returns exponentially to the synchronization manifold [Fig. 6.1(a)], if it is unstable the trajectory will diverge exponentially from the synchronization manifold [Fig. 6.1(b)].

Given the solution of the equations of motion of the coupled oscillators provided by Eq. 6.3, which describe a synchronized evolution in the synchronization manifold, if a perturbation brings the system state out of the manifold, the motion in phase space will be given by $\mathbf{x}_1(t) = \mathbf{x}(t) + \delta\mathbf{x}_1(t)$ and $\mathbf{x}_2(t) = \mathbf{x}(t) + \delta\mathbf{x}_2(t)$, which verify Eqs. 6.1–6.2. The perturbations $\delta\mathbf{x}_1(t)$ and $\delta\mathbf{x}_2(t)$ then, up to first order in $\delta\mathbf{x}$, verify

$$\frac{d\delta\mathbf{x}_1}{dt} = \left[\frac{\partial\mathbf{F}(\mathbf{x})}{\partial\mathbf{x}}\right]_{\mathbf{x}(t)} \cdot \delta\mathbf{x}_1 + \mathbf{C}\cdot(\delta\mathbf{x}_2 - \delta\mathbf{x}_1), \tag{6.4}$$

$$\frac{d\delta\mathbf{x}_2}{dt} = \left[\frac{\partial\mathbf{F}(\mathbf{x})}{\partial\mathbf{x}}\right]_{\mathbf{x}(t)} \delta\mathbf{x}_2 \cdot + \mathbf{C}\cdot(\delta\mathbf{x}_1 - \delta\mathbf{x}_2). \tag{6.5}$$

Instead of looking to the evolution of the projections of the perturbation onto the $\mathbf{x}_1$ and $\mathbf{x}_2$ subspaces, it is useful to work with the projections onto the synchronization manifold, $\delta\mathbf{x}_{\parallel}(t)$, and onto the transverse subspace, $\delta\mathbf{x}_{\perp}(t)$. These are given by $\delta\mathbf{x}_{\parallel} = (\delta\mathbf{x}_1 + \delta\mathbf{x}_2)/\sqrt{2}$ and $\delta\mathbf{x}_{\perp} = (\delta\mathbf{x}_1 - \delta\mathbf{x}_2)/\sqrt{2}$, and are called the parallel perturbation, and the transverse perturbation, respectively. Their time evolution is then obtained from Eqs. 6.4–6.5 written in terms of these new projections

$$\frac{d\delta\mathbf{x}_{\parallel}}{dt} = \left[\frac{\partial\mathbf{F}(\mathbf{x})}{\partial\mathbf{x}}\right]_{\mathbf{x}(t)} \cdot \delta\mathbf{x}_{\parallel}, \tag{6.6}$$

$$\frac{d\delta\mathbf{x}_{\perp}}{dt} = \left[\frac{\partial\mathbf{F}(\mathbf{x})}{\partial\mathbf{x}} - 2\mathbf{C}\right]_{\mathbf{x}(t)} \cdot \delta\mathbf{x}_{\perp}, \tag{6.7}$$

which are two independent equations that have the same structure that the linearized equation (Eq. 2.21), which is the base of the calculation of Lyapunov exponents. The condition of stability is then easily formulated by saying that the Lyapunov exponents for perturbations perpendicular to the synchronization manifold have to vanish exponentially; this means that the Lyapunov exponents obtained from the variational equation of the transverse part of the perturbation (Eq. 6.7) have to be negative. It is to be noted that Eq. 6.7 has the Jacobian matrix, $(\partial \mathbf{F}/\partial \mathbf{x})_{\mathbf{x}(t)}$, substituted by $(\partial \mathbf{F}/\partial \mathbf{x})_{\mathbf{x}(t)} - 2\mathbf{C}$; therefore, these exponents, which are called transverse Lyapunov exponents, will depend on the numerical values of the components of the matrix $\mathbf{C}$. For a given system, the calculation of these exponents as functions of $\mathbf{C}$ will determine when the two oscillators allow asymptotically stable synchronization. Because of the identical structure of Eq. 6.7 and Eq. 2.21 the techniques for calculation of Lyapunov spectra presented in Chapter 2 can be adapted straightforwardly to compute the transverse Lyapunov exponents.

All this will now be illustrated by means of the example of two identical Lorenz oscillators [Lorenz (1963)] mutually coupled. The six-dimensional system of equations of motion is

$$dx_{1,2}/dt = \sigma\,(y_{1,2} - x_{1,2}) + C\,(x_{2,1} - x_{1,2})\,, \tag{6.8}$$

$$dy_{1,2}/dt = x_1\,(r - z_{1,2}) - y_{1,2} + C\,(y_{2,1} - y_{1,2})\,, \tag{6.9}$$

$$dz_{1,2}/dt = x_{1,2}y_{1,2} - bz_{1,2} + C\,(z_{2,1} - z_{1,2})\,, \tag{6.10}$$

where it has been assumed that the interaction matrix is of the form $\mathbf{C} = C\mathbf{I}$, with $\mathbf{I}$ the identity matrix and C a scalar constant. The matrix to be used in Eq. 6.7 to analyze the stability of the synchronized state is

$$\frac{\partial \mathbf{F}(\mathbf{x})}{\partial \mathbf{x}} - 2\mathbf{C} = \begin{bmatrix} -(\sigma + 2C) & \sigma & 0 \\ r - z & -(1 + 2C) & -x \\ y & x & -(b + 2C) \end{bmatrix}. \tag{6.11}$$

For this example, the parameters of the Lorenz system have been fixed in $\sigma = 10$, $r = 60$ and $b = 8/3$, which correspond to a chaotic state. The three transverse Lyapunov exponents, computed as functions of the strength of the coupling, C, are presented in Figs. 6.2(a, b): they decrease monotonously, from the values of the Lyapunov exponents of a single free Lorenz oscillator, with increasing coupling strength. The positive exponent becomes negative at the transition value $C_T \approx 0.71$ which signals the onset of the synchronized motion.

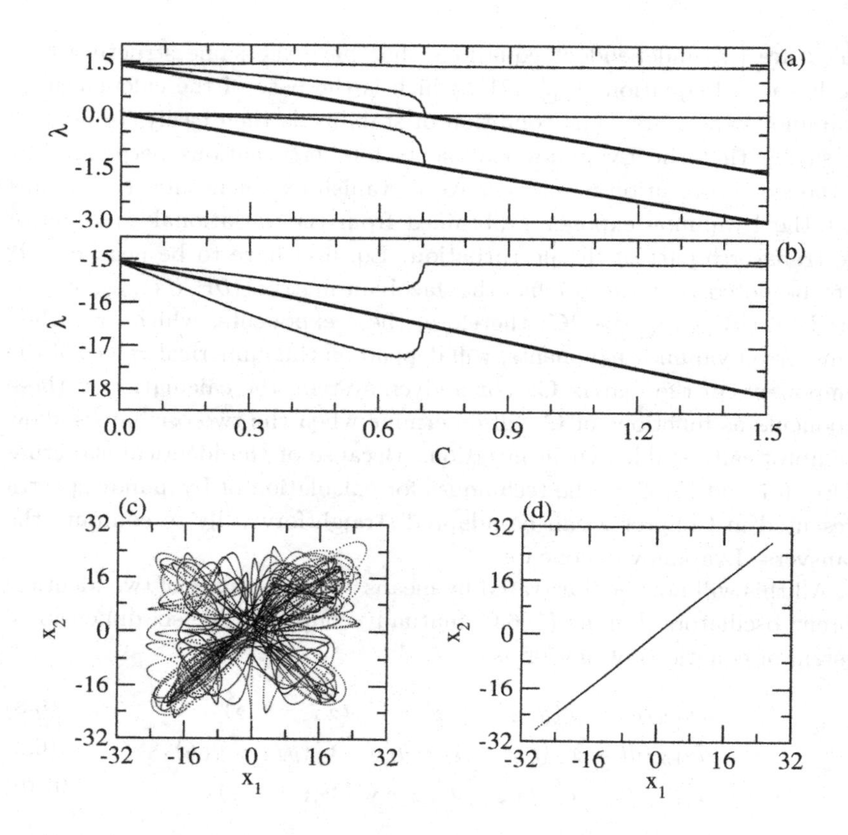

Fig. 6.2 Stability analysis of a system of two Lorenz oscillators mutually coupled :(a) two largest transverse Lyapunov exponents (thick lines) and four largest Lyapunov exponents of the six-dimensional system (thin lines), and (b) third transverse Lyapunov exponent (thick line) and fifth and sixth Lyapunov exponents (thin lines), all of them as functions of the coupling strength. Parametric plots, $x_2(x_1)$, (c) below ($C = 0.6$), and (d) above ($C = 0.8$) the transition to synchronization.

The dependence of the spectrum of Lyapunov exponents on C for the six-dimensional system of Eqs. 6.8–6.10 has also been plotted in Figs. 6.2(a, b). The first exponent stays practically constant and positive, $\lambda_1 \approx 0.14$, in the whole range of values of C studied; this means that the coupled system stays chaotic even in the asymptotically stable synchronized state. In fact, it is hyperchaotic below C_T, and the transition from hyperchaos to chaos at C_T, is also a signal of the transition to stable synchronization. Above C_T the motion is restricted to the synchronization manifold, and the Lyapunov exponents are coincident with the longitudinal and transverse exponents. Three Lyapunov exponents (λ_1, λ_2, and λ_4) describe the motion in the synchronization manifold and are equal to the longitudinal Lyapunov exponents ($\lambda_1^{\parallel}$, $\lambda_2^{\parallel}$, and $\lambda_3^{\parallel}$), which have the same values as the Lyapunov exponents of an independent oscillator: $\lambda_1 = \lambda_1^{\parallel} \approx 1.4$, $\lambda_2 = \lambda_2^{\parallel} \approx 0$, and $\lambda_4 = \lambda_3^{\parallel} \approx -15$. The other three Lyapunov exponents (λ_3, λ_5, and λ_6) describe the motion in the transverse subspace and are equal to the transverse Lyapunov exponents ($\lambda_1^{\perp}$, $\lambda_2^{\perp}$, and $\lambda_3^{\perp}$); i.e. $\lambda_3 = \lambda_1^{\perp}$, $\lambda_5 = \lambda_2^{\perp}$, and $\lambda_6 = \lambda_3^{\perp}$; because the stable motion is limited to the synchronization manifold these are all negative.

Additional illustration of this phenomenon of identical synchronization between mutually coupled chaotic oscillators is provided in Figs. 6.2(c, d) which show parametric plots of $x_2(t)$ versus $x_1(t)$, below and above C_T, respectively. These have been obtained from trajectories computed from initial conditions that are outside of the synchronization manifold. The cloudy distribution of points in Fig. 6.2(c) indicates a desynchronized state, and the 45° segment in Fig. 6.2(d) is the well known sign of identical synchronization.

6.1.2 *Complexity in the transition to the synchronized state*

The transition from desynchronization to synchronization presented in the above section is simple, clean and smooth; this, however, is not the general case, and even a system so structurally simple as a set of two identical mutually coupled chaotic oscillators is a complex system which allows a rich variety of dynamical behaviors. Besides the synchronized and the desynchronized states, exemplified in Fig. 6.2, it is possible to observe other kinds of solutions of Eqs. 6.1–6.2. Usually as the coupling strength is increased from zero, there is a desynchronized state for weak coupling, and synchronization for sufficiently large intensity of the coupling. In the middle, in many systems, it is possible to observe a set of other different dynamical behaviors that make the transition to the synchronized state a complex process.

Two main sources of complexity arise: changes of the chaotic nature of the dynamics, and multistability. As the coupling strength increases, there can be chaos–chaos transitions in which the nature of the dynamics is altered in some sense but still remaining chaotic; a particular example of this kind of transition is the interior crisis [Grebogi et al. (1982); González-Miranda (2003)]. There can also be transitions from chaoticity to periodicity; frequently by means of a inverse period doubling cascade, or through an intermediate quasiperiodic attractor. Moreover, the final state achieved by the coupled oscillators, for a given coupling strength, may be different for different initial conditions of the oscillators; this is what is called multistability [Grebogi et al. (1983)].

One example of complex transition is now presented by means of two mutually coupled Rössler oscillators [Rössler (1976)] given by

$$dx_{1,2}/dt = -\left(y_{1,2} + z_{1,2}\right) + C\left(x_{2,1} - x_{1,2}\right), \tag{6.12}$$

$$dy_{1,2}/dt = x_{1,2} + ay_{1,2} + C\left(y_{2,1} - y_{1,2}\right), \tag{6.13}$$

$$dz_{1,2}/dt = b + z_{1,2}(x_{1,2} - c) + C\left(z_{2,1} - z_{1,2}\right), \tag{6.14}$$

with an interaction matrix of the form $\mathbf{C} = C\mathbf{I}$ as above. A study of the transitions between desynchronization and synchronization for the system parameter values $a = 0.2$, $b = 0.2$, and $c = 5.7$ is presented in Fig. 6.3 and in Fig. 6.4.

The dynamical states achieved by the system as functions of the coupling strength, C, are characterized by means of bifurcation diagrams obtained numerically from the set of values of the maxima reached by the x signal of each system, and by means of the spectra of Lyapunov exponents of the six-dimensional system. All these are computed from trajectories obtained from Eqs. 6.12–6.14 for a set of 500 values of C, starting in $C = 0$ and progressively increasing the value of C by small steps ΔC [Figs. 6.3 (a, b)]. Otherwise [Figs. 6.3 (c, d)] the calculation was started at $C = 0.05$ and the value of C was decreased by steps ΔC . The initial conditions chosen for $C = 0$ in the first case, or for $C = 0.05$ in the second, were arbitrary phase space points of the attractors of the isolated oscillators. For all the other values of C the coordinates of the last point of the trajectory of the previous run, randomly modified by a one percent were used as initial conditions for that run. Moreover, the synchronization condition has been tested by means of the calculation of the transverse Lyapunov exponents.

The bifurcation diagrams obtained are the same for the two oscillators. One of them, presented in Fig. 6.3 (a), displays a broad band for small values of the coupling strength ($C \lesssim 0.017$), and for large values ($C \gtrsim 0.045$). The first band corresponds to a regular desynchronized hyperchaotic attractor, and the second to identical chaotic synchronization. The chaotic behavior can be inferred from the band structure of the bifurcation diagram, and from the three largest Lyapunov exponents presented in Fig. 6.3(b), were it is seen that two of them are positive in the first region, and only one is positive in the second. The synchronization condition is tested by means of

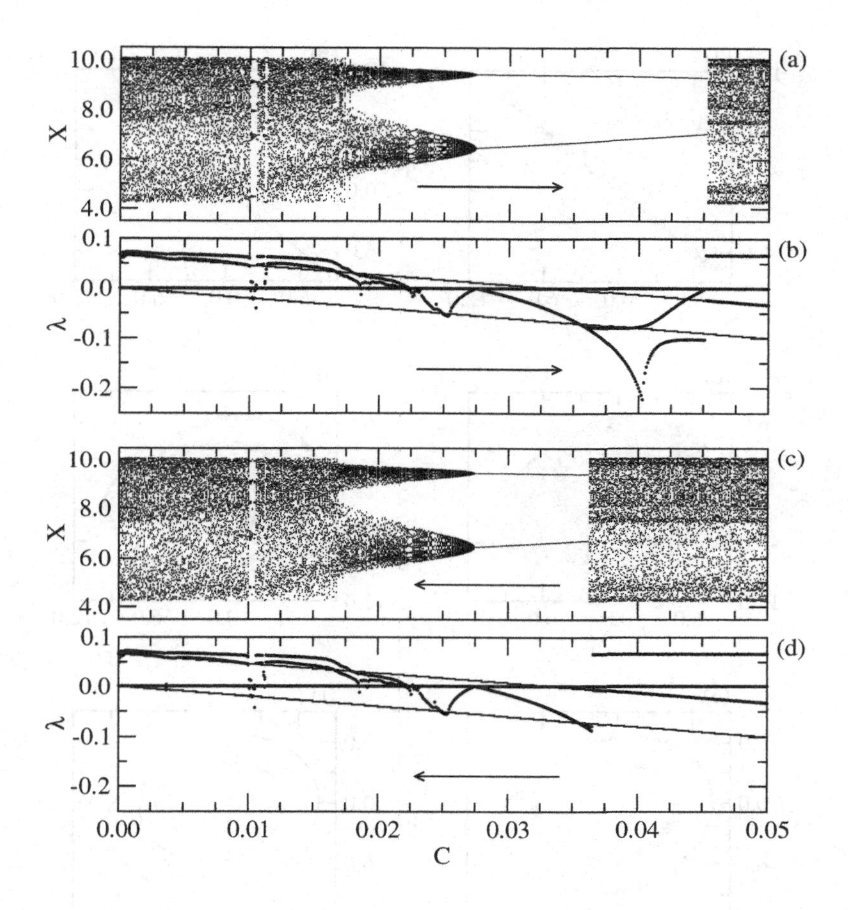

Fig. 6.3 Study of the transition between the synchronized and the desynchronized regimes of two identical Rössler oscillators mutually coupled. A transition from the desynchronized state to the synchronized state is shown by means of (a) the bifurcation diagram of system 1, obtained from the returns of the x-signal, and (b) the three largest Lyapunov exponents (thick lines), and the two largest transverse Lyapunov exponents (thin lines). Plots (c) and (d) are the same as (a) and (b) respectively, but for a transition from the synchronized state to the desynchronized state. The arrows in all these plots signal the direction of the transition in each case.

the largest transverse Lyapunov exponent, which is positive for $C \lesssim 0.017$, and negative for $C \gtrsim 0.045$. Additional illustration of the synchronization condition is given by parametric plots of x_2 versus x_1, which show a cloudy structure for $C = 0.003$ in Fig. 6.4(a), and a clean straight line segment for $C = 0.047$ in Fig. 6.4(b).

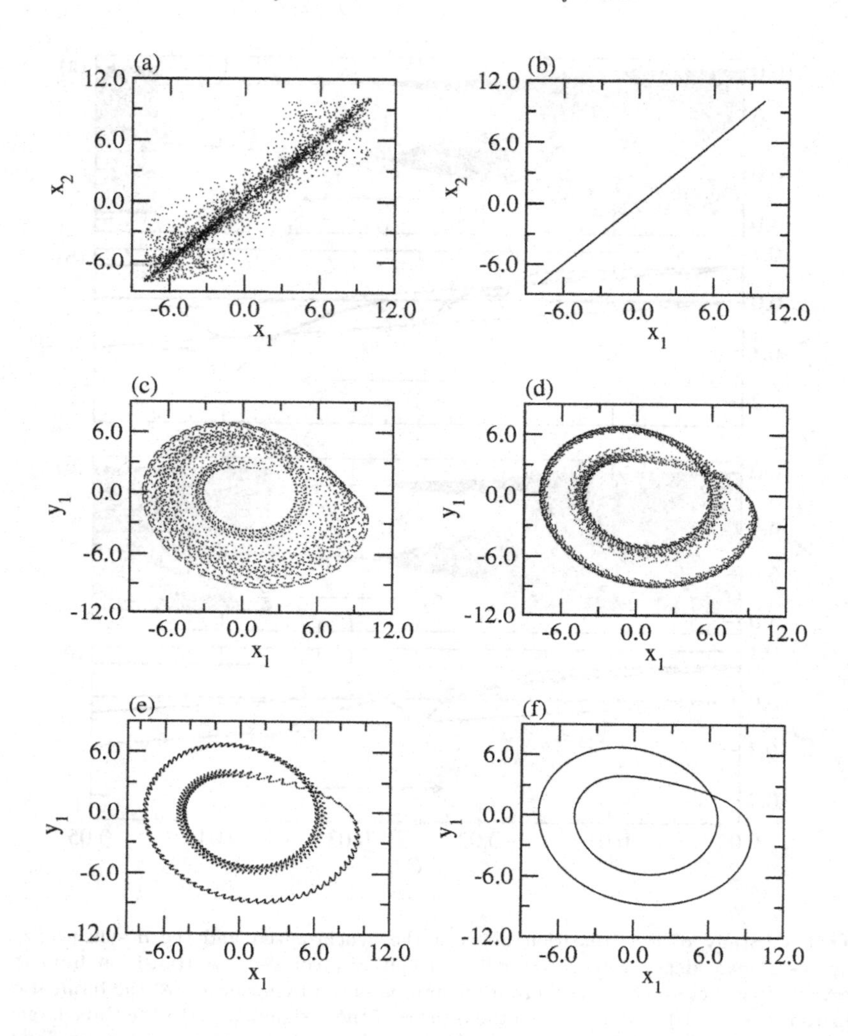

Fig. 6.4 Dynamic behaviors available to two coupled Rössler oscillators as shown by parametric plots for the x-signals for (a) a desynchronized state ($C = 0.003$), and (b) a synchronized state ($C = 0.047$), and by the several types of attractors allowed to each oscillator: (c) one band chaos ($C = 0.003$), (d) two bands chaos ($C = 0.020$), (e) quasiperiodicity ($C = 0.025$), and (f) periodicity ($C = 0.035$).

For values of the coupling strength in the intermediate region ($0.017 \lesssim C \lesssim 0.045$) the bifurcation diagram in Fig. 6.3(a) does not present a simple band structure anymore, but a variety of structures which are signals of complexity in the transitions between the two regular states discussed in the above paragraph. The bifurcation diagram and the Lyapunov spectra indicate the existence of three qualitatively different regimes in this region.

In the interval $0.017 \lesssim C \lesssim 0.022$ the system stays hyperchaotic, but the structure of the bifurcation diagram has abruptly changed from a single band to a double band; moreover, the values of the two positive Lyapunov exponents have notably decreased. This is an example of chaos–chaos transition which is further illustrated in Fig. 6.4(c) and Fig. 6.4(d) which show the projections of the trajectories of the first oscillator in the $x - y$ plane for $C = 0.003$ and for $C = 0.020$, respectively: the change in the topology is evident, a single band attractor becomes a double band attractor.

In the interval $0.022 \lesssim C \lesssim 0.0275$, the two largest Lyapunov exponents have become null, which now means that the motion is not chaotic but quasiperiodic. The trajectories of each oscillator still have the structure of a double band [Fig. 6.4(e)].

For $0.0275 \lesssim C \lesssim 0.045$ the bifurcation diagram is made of two lines, and of the two largest Lyapunov exponents one is null and the other negative; therefore, the motion occurs in a limit cycle as further illustrated in Fig. 6.4(f). Furthermore, in this interval the largest transverse Lyapunov exponent becomes negative for $C_T \approx 0.033$, so there are two different regimes of periodicity: one desynchronized below C_T, and one synchronized above C_T. Finally, for $C \approx 0.045$, a discontinuous transition occurs that makes the synchronized state chaotic. This presents the qualitative features of an exterior crisis [Grebogi et al. (1982)].

This example is also useful to illustrate an additional complex behavior: multistability. This is demonstrated in Figs. 6.3 (c, d) where the bifurcation diagram and Lyapunov exponents have been computed for a transition from the synchronized to the desynchronized states. The dynamics observed in the interval $0.036 \lesssim C \lesssim 0.045$ is different to the one observed in Figs. 6.3 (a, b) for the same interval. This means that, within this interval, different dynamical behaviors coexist for the coupled oscillators, that can be attained from different basins of attraction in the space of initial conditions.

It has to be noted that the details of the transition presented in Figs. 6.3 and 6.4 are not universal, but just an illustration of the kind of complexity that may be expected in the transition from desynchronization to synchronization.

6.1.3 *Numerical simulations and experiments*

The Lorenz and the Rössler models have been used to perform numerical studies of the synchronization phenomena described above. The configuration presented here for two Lorenz oscillators mutually coupled, although using other parameter values ($\sigma = 10$, $r = 28$, and $b = 8/3$), was studied numerically [Fujisaka and Yamada (1983)], using a somewhat different algorithm to compute the first Lyapunov exponent and the first transverse Lyapunov exponents, to illustrate the main properties of the synchronization of coupled chaotic oscillators. A study [Pecora et al. (1997)] of two Rössler oscillators mutually coupled only through the first variable ($C_{x,x} = C$, and $C_{\alpha,\beta} = 0$ otherwise) showed that the behavior of $\lambda_1^{\perp}$ for values of C within the range studied here is similar to that presented for the Lorenz system: a monotonous decay with a transition point, C_T, where the transverse exponent becomes negative. The study was extended to large values of the coupling strength ($C \gtrsim 5C_T$), where it was found that the exponent became positive and stable synchronization was lost.

The synchronization of identical systems mutually coupled, as those modelled by Eqs. 6.1–6.2 has been demonstrated experimentally by means of chaotic electric circuits. The realization of mutual coupling is quite simple because it can be done by connecting two appropriate equivalent nodes of each circuit with an electric resistance whose value, R, controls the strength of the coupling. For example, for two coupled Chua circuits [Matsumoto et al. (1985)], labeled 1 and 2, the realization of the following coupling scheme

$$dx_{1,2}/dt = \alpha\left[y_{1,2} - x_{1,2} - f(x_{1,2})\right], \tag{6.15}$$

$$dy_{1,2}/dt = x_{1,2} - y_{1,2} + z_{1,2} + C(y_{2,1} - y_{1,2}), \tag{6.16}$$

$$dz_{1,2}/dt = -\beta y_{1,2} \tag{6.17}$$

can be realized by means of an experimental set up like the one presented in Fig. 4.6 with the only change of taking off the operational amplifier in the middle of the figure and leaving the two circuits connected just by the adjustable electrical resistor R_C. The intensity of the coupling will then be given by $C \propto 1/R_C$. Experiments with this and other similar configurations have indeed been performed [Chua et al. (1992)]. These allowed to observe synchronized states by means of parametric plots for coupling strengths that are large enough.

Phenomena of chaos synchronization with a configuration of this type, but using a different electric circuit [Afraimovich et al. (1986)], have also been reported [Rulkov et al. (1992)]. These authors observed chaos synchronization above certain coupling strength, and substantiated their experiments with numerical studies of the stability of the synchronization similar to those presented in the above subsection. Moreover, the structural stability of the synchronization was proven by considering small differences between the coupled circuits which caused few changes in the dynamical behavior.

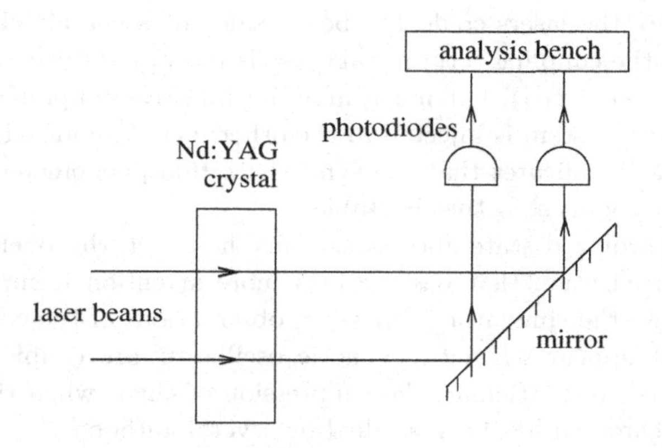

Fig. 6.5 Basic elements of an experiment designed to study the dynamics of mutually coupled chaotic lasers. The laser is a Nd:YAG crystal pumped by a modulated beam coming from another laser (an Ar laser, for example). Two nearly identical lasers can be generated by pumping different regions of the same crystal by two identical parallel laser beams created from an original single beam by means of a beam splitter (See Fig. 4.7). If the two pumping beams are close enough, the Nd:YAG lasers formed are mutually coupled by means of the overlap of their fields.

Stable synchronization in experiments made with two mutually coupled lasers has also been reported [Roy and Thornburg (1994)]. The chaotic oscillators were made by a single crystal composed of a Neodymium doped Yttrium Aluminum garnet ($Nd : YAG$) as illustrated in Fig. 6.5. The crystal was pumped by two parallel beams obtained by splitting a single beam generated by an Argon laser. The two beams act on regions of the crystal separated a distance, D, large compared to the size of the cross section of the beams. The interaction of each beam with the crystal generates a single local $Nd : YAG$ laser which, in the conditions of the experiment, displays chaotic intensity fluctuations. The coupling between these two oscillators is made by means of the overlap of the laser fields in the crystal, and can be controlled by changing the distance between the beams. The output of each of the two $Nd : YAG$ lasers is sent to a photodiode, and then to a digital oscilloscope for analysis. By decreasing D from values where the two lasers oscillated independently it was possible to arrive to values of the coupling strong enough to observe synchronized fluctuations of the laser intensities by means of both time series and parametric plots [Roy and Thornburg (1994)]. This experiment showed the possibility of synchronizing two lasers mutually coupled. Moreover, because due to imperfections in the experimental set up, the lasers could not be considered as completely identical, and because the coupling term in this case is not of a diffusive type, modeled by $C \cdot (x_{2,1} - x_{1,2})$, but much more an injective coupling in which a signal from one system is injected in the other, better modeled as $C \cdot x_{2,1}$, this experiment indicates that the synchronization phenomena in coupled chaotic oscillators is structurally stable.

The synchronized state and its stability has been the phenomenon of chaos synchronization that has received more attention from the experimentalists and the simulators. However, observations of the complex phenomena that appear when two chaotic oscillators are coupled have not been neglected; in particular, the suppression of chaos when the coupling strength is increased has been studied by several authors.

Experimental reports of chaos suppression after coupling chaotic electric circuits have been made. In the experiment, discussed above [Rulkov et al. (1992)], of two identical chaotic electric circuits coupled diffusively, the observation of the development of limit cycles after the increase of the coupling strength were also mentioned. In later investigations of the same electric circuit [del Rio et al. (1994)], the observation of the suppression of chaos with increasing coupling strength by means of an inverse period doubling cascade was reported. After further increase of the coupling strength a chaotic synchronized state suddenly developed in what appears to be a scenario similar to that described in the above subsection.

A somewhat different phenomenon of chaos suppression has been observed in numerical simulations of a model of two identical mutually coupled semiconductor lasers [Bindu and Nandakumaran (2000)]. The relevant variables for each laser are the output photon density, and the carrier density; moreover there is the driving current which pumps each laser. The coupling was made by adding a current proportional to the output photon density of each laser to the pumping driving current. These authors observed, by means of parametric plots of the output photon density, that as the coupling strength increases the outputs of the lasers synchronize when a certain threshold value is overcome. Further increase of the coupling results in the destruction of chaos in the two lasers by means of an inverse period doubling cascade. This was seen by means of phase space plots of the attractors, whose structure evolved from a fully chaotic structure to complex periodic orbits, to end in a simple limit cycle that lasts for very large values of the coupling strength.

The experimental observation of the suppression of chaos after coupling two CO_2 lasers by sending an small part of the output of each laser to the other has been reported [Liu et al. (1994)] in an experimental set similar to that presented in Fig. 4.7, with the optical insulator removed to have mutual coupling. Parametric plots, and time series of the intensity of their output beams showed the destruction of chaos after coupling.

6.2 Dynamics of systems of $N > 2$ coupled oscillators

Systems made of more than two coupled chaotic oscillators making some sort of network are useful to study the dynamics of physical, chemical and biological systems that have many degrees of freedom. These can be more or less complicated networks of interconnected elementary units, or continuous extended systems that are modelled by means of partial differential equations. In this last case the numerical treatment of these equations may result in a reformulation of the mathematical problem in terms of networks of coupled chaotic oscillators. Collective behaviors such as turbulence, spatial variations of observables, pattern formation, and clustering are the kind of phenomena that are of interest in this context. A desynchronized state is a model of turbulence in space an time, while a synchronized chaotic state describes spatially ordered and temporary chaotic, or turbulent, systems. Intermediate states in which only groups of oscillators synchronize correspond to phenomena of pattern formation and clustering.

6.2.1 *Basic theory*

Besides the simple case of two coupled chaotic oscillators defined by Eqs. 6.1–6.2, the study of the dynamics of collectives made of a number $N > 2$ of oscillators coupled along different configurations has received considerable attention in the literature.

In these cases, given the individual chaotic flow $d\mathbf{x}/dt = \mathbf{F}(\mathbf{x})$, the systems of interest used to be well described by

$$\frac{d\mathbf{x}_i}{dt} = \mathbf{F}(\mathbf{x}_i) + C \cdot \sum_{j=1}^{N} c_{i,j}\left(\mathbf{x}_j - \mathbf{x}_i\right),\ i = 1, 2, ..., N, \tag{6.18}$$

with C a scalar constant which gives an overall measure of the strength of the interaction between the N oscillators, and $c_{i,j}$ a matrix which describes the configuration of the interactions, and is known as the coupling matrix. This is frequently chosen [Fujisaka and Yamada (1983)] such that the following symmetry rule

$$\sum_{j=1}^{N}\sum_{j=1}^{N} c_{i,j}\left(\mathbf{x}_j - \mathbf{x}_i\right) = 0 \tag{6.19}$$

is verified.

There are many choices of $c_{i,j}$ that make sense. Two configurations very frequently found are: (i) $c_{i,j} = 1$ for all $i, j \in N$, which describes an identical all to all interaction, also known as global coupling, (ii) $c_{i,j} = 1$ if $|i - j| = 1$, and $c_{i,j} = 0$ otherwise. This describes a linear chain with interactions only to nearest neighbors, which is called diffusive coupling. More sophisticated configurations can be made by choosing the values of $c_{i,j}$ different for each pair to describe specific spatial arrangements of the oscillators, as well as special types of interaction. Approximations for the actions on the i-th oscillator that deviate from $C \cdot \sum_{j=1}^{N} c_{i,j}\left(\mathbf{x}_j - \mathbf{x}_i\right)$, have also been used; one example is the mean field approximation, in which an all to all interaction is averaged in the form

$$C \cdot \sum_{j=1}^{N} c_{i,j}\left(\mathbf{x}_j - \mathbf{x}_i\right) \approx \frac{C}{N} \cdot \sum_{j=1}^{N} \mathbf{x}_j \tag{6.20}$$

Moreover, it deserves to be noted that Eq. 6.18 can be generalized to study spatiotemporal phenomena just by using a non-autonomous field, $\mathbf{F}(\mathbf{x}_i, t)$, for the dynamics of each individual oscillator.

In all of the above it is assumed that the system of oscillators is isolated. In some applications, such as the study of continuous systems, or of specific experimental configurations found in practice, boundary conditions appropriate to the case studied have to be used. A case frequently found in the literature is the use of periodic boundary conditions in which the finite system of N oscillators is assumed to be surrounded by a set of identical copies in a number and spatial arrangement given by the structure of the coupling matrix, $c_{i,j}$. For example, for a linear chain of N oscillators with only nearest neighbor interactions, where the coupling term is given by $C \cdot \sum_{j=1}^{N} c_{i,j} (\mathbf{x}_j - \mathbf{x}_i) = C \cdot (\mathbf{x}_{i-1} + \mathbf{x}_{i+1} - \mathbf{2x}_i)$, the periodic boundary conditions are written as $\mathbf{x}_{N+1}(t) = \mathbf{x}_1(t)$. This particular system is equivalent to a closed ring of oscillators, a geometrical configuration frequently found in experiments.

The system of equations Eq. 6.18 always allows the solution

$$\mathbf{x}_1(t) = \mathbf{x}_2(t) = ... = \mathbf{x}_N(t) = \mathbf{x}(t). \tag{6.21}$$

when $\mathbf{x}(t)$ is a solution of differential equation $d\mathbf{x}/\mathbf{dt} = \mathbf{F}(\mathbf{x})$. A state like this represents a very coherent form of motion in which all oscillators although chaotic are synchronized in the sense that there is identical synchronization between any pair of oscillators of the system. This is a generalization of the state defined by Eq. 6.3 for just two oscillators, and is also called the macro-oscillation or synchronized state [Fujisaka and Yamada (1983)].

As before, the fundamental question here is the stability of the synchronized state. All the stability theory developed above, formulated in terms of the synchronization manifold and the transverse subspace, and developed by means of the study time evolution of perturbations as given by a linearization of the equations of motion can be generalized to this case [Fujisaka and Yamada (1983); Pecora and Carroll (1998)]. The condition of the stability of the synchronized state that is then obtained is the same as before: all the transverse Lyapunov exponents, i.e. the Lyapunov exponents for perturbations perpendicular to the synchronization manifold, have to be negative to the synchronized state to be stable [Fujisaka and Yamada (1983)]. Moreover, based on these concepts, there have been developed [Pecora and Carroll (1998)] techniques to compute stability diagrams which, for a given system made of N oscillators and a given coupling scheme, allow the prediction of the conditions under which there will be stable synchronization.

6.2.2 *Partial synchronization of chaos*

When a set of $N > 2$ chaotic oscillators are mutually coupled they may display either a desynchronized state, for weak coupling, in which all the oscillators evolve in chaotic trajectories uncorrelated to each other; or a synchronized state, for a larger coupling strength, in which all oscillators remain chaotic with their trajectories synchronized in the sense that Eq. 6.21

is verified. Moreover, for intermediate coupling, it is possible to observe complexity, including multistability and transitions between different types of chaotic and non-chaotic dynamics. All this phenomenology is essentially the same described in subsection 6.1.2 for only two coupled oscillators. Moreover, when more than two oscillators are coupled a new phenomenology which is called partial synchronization of chaos [Kaneko (1990); Hu et al. (2000)] is possible. This is defined as the case when, some but not all the oscillators synchronize to each other. When partial synchronization is achieved the system of N oscillators can be decomposed in a certain number of subsystems, not necessarily of the same size, such that all oscillators in each subsystem are in synchrony to each other, but desynchronized from all the other oscillators. This is a very interesting state because it provides a mechanism for pattern formation and clustering in a set of coupled chaotic oscillators.

A state of partial synchronization of chaos will be demonstrated here by means of the example of $N = 6$ Rössler oscillators coupled to nearest neighbors in a linear chain

$$dx_i/dt = -\left(y_{1,2} + z_{1,2}\right) + C\left(x_{i-1} + x_{i+1} - 2x_i\right), \tag{6.22}$$

$$dy_i/dt = x_{1,2} + ay_{1,2} + C\left(y_{i-1} + y_{i+1} - 2y_i\right), \tag{6.23}$$

$$dz_i/dt = b + z_{1,2}(x_{1,2} - c) + C\left(z_{i-1} + z_{i+1} - 2y_i\right), \tag{6.24}$$

with $i = 1, 2, ..., 6$, and subject to periodic boundary conditions so that $x_{N+1} = x_1$, $y_{N+1} = y_1$, and $z_{N+1} = z_1$. This is equivalent to considering a closed ring of $N = 6$ oscillators. As in the example of two oscillators presented in Section 6.1, the parameters for each individual oscillator have been fixed in $a = 0.2$, $b = 0.2$, and $c = 5.7$.

For small and large coupling the regular desynchronized and synchronized states can be observed. At intermediate coupling several forms of complexity develop, among them partial synchronization of chaos. This last is illustrated in Fig. 6.6 for a coupling strength $C = 0.052$. Parametric plots obtained from the x variable for all possible couples of oscillators, presented in Fig. 6.6(a), show cloudy sets of points for all couples except for oscillators 1 and 3, and for oscillators 4 and 6, which display the characteristic straight line segments of identical synchronization. All oscillators remain chaotic in this state, for example trajectory plots for two of them are shown in Figs. 6.6(b, c); they display a structure similar to that of the independent oscillators. The system states that results are presented in a schematic form in Figs. 6.6(d) where each oscillator is represented by a

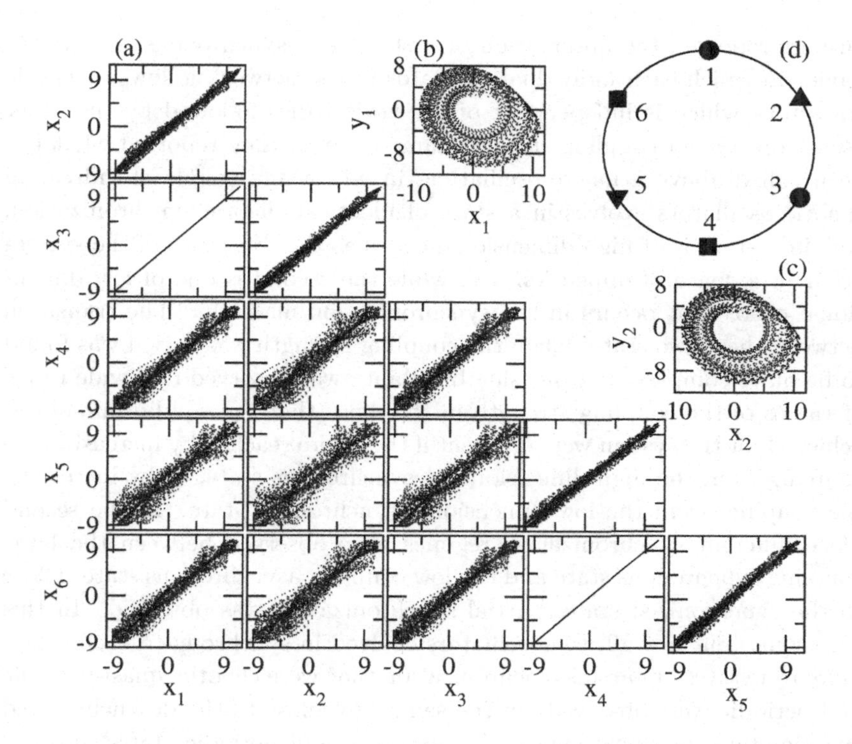

Fig. 6.6 (a) A state of partial synchronization of chaos in a set of six identical Rössler oscillators coupled to nearest neighbors ($C = 0.052$) in a linear chain with periodic boundary conditions demonstrated by means of the 15 different parametric plots available for the x-signals of the oscillators. Projections of the trajectories in the $x - y$ plane for oscillators (b) number 1, and (c) number 2. (d) Schematic plot of the chain of oscillators, presented as a ring because of the boundary conditions, with each oscillator represented by a different symbol according to its state of synchronization (mutually synchronized oscillators share the same symbol).

symbol (circle, square, and one of two types of triangles) connected by an arch to its nearest neighbors; oscillators following identical trajectories are represented by the same symbol. The particular pattern attained is obvious in this plot.

6.2.3 *Computer simulations and laboratory experiments*

The dynamics of chains of N Rössler oscillators with diffusive coupling to nearest neighbors, periodic boundary conditions, and $5 \leq N \leq 50$ has been studied by computer simulation [Hu et al. (2000)]. For weak coupling, these authors reported the observation of a state of desynchronous chaotic behavior in which each individual oscillator in the network follows a chaotic trajectory which is independent of the trajectories followed by the other oscillators. For a coupling strength large enough, they reported what has been called above a macro-oscillation, in which any couple of individual chaotic oscillators evolves in a state of identical chaotic synchronization. The first state is of high dimensional chaos as the dynamics of the system occur in a space of dimension $3N$, while the second is one of low dimensional chaos that occurs in the synchronization manifold. The transition between these two states when the coupling strength was varied was found to be quite complex. For one side bistability was observed in a wide range of values of the coupling strength in the sense that the equilibrium states achieved for the system were different if they were reached by increasing the coupling from the high dimensional desynchronous state, or by increasing the coupling from the low dimensional synchronous state. In the second place different synchronization regimes were observed between the large coupling synchronous state and the low coupling asynchronous state. Close to the synchronized state, partial synchronization was observed. In this case some, but not all, the oscillators evolved in synchrony to each other; moreover states of partial synchronization that were chaotic, quasi-periodic and periodic were observed. In the same way, close to the desynchronized chaotic state, desynchronized quasi-periodic and periodic states also appeared.

Experiments aimed to observe chaos synchronization in a ring of $N = 4$ nearly identical electric circuits, diffusively coupled to nearest neighbors have been performed [Heagy et al. (1994)]. The circuit, called a modified Rössler system, was designed to mimic the behavior of the Rössler attractor. The coupling was made through auxiliary circuits which combine addition of signals by means of operational amplifiers, to create a signal of the type $(s_{i+1} + s_{i-1} - 2s_i)$ from the scalar signals s of neighbor circuits, and analog multipliers to multiply this signal by a constant to control the coupling strength. The experiments were made coupling through just one variable. Three cases were considered: coupling through x, through y, and through z. In the first two cases, synchronized states were reported to occur above certain threshold value of the coupling strength; these were observed between parametric plots of three couples of variables. No synchronization was observed for z-coupling. Numerical analysis in terms of Lyapunov exponents, performed from the equations of the systems, were found in agreement with the experimental observations.

Rings of $N = 6$ of the above nearly identical modified Rössler systems diffusively coupled have also been used to experimentally demonstrate partial synchronization [Heisler et al. (2003)]. These authors also reported the observation of a homogeneous synchronized state when the coupling was made by only one scalar signal. For coupling through two or three scalar signals they obtained partial synchronization of chaos in which the systems of six oscillators could be divided in two or three subsystems, of three oscillators, or two oscillators each. The individual oscillators in each subsystem evolved in synchrony between them, while they where desynchronized with the oscillators of the other subsystems.

Another experimental system that has been used to study the synchronization of systems of $N > 2$ coupled chaotic oscillators is an electrochemical cell with the working electrode substituted by an array made up of N individual electrodes [Wang et al. (2000)]. The array consists of 64 wires isolated in such a way that chemical reactions take place only in the tips of each wire. Each of these 64 chemical reactions constitutes one individual oscillator of the system of N nearly identical chaotic chemical oscillators that are under study. Each electrode is connected to an individual resistor; and, there is a collective resistor which connects all these individual resistor electrodes to a potentiostat which maintains the electrodes at the same potential. This form of connection creates a global coupling whose strength can be controlled. The individual currents at each electrode can be measured, and used to reconstruct the three-dimensional dynamics of each oscillator by means of embedding techniques. This allowed to monitor the dynamics of the systems by following the dynamics of each individual oscillator in the same three-dimensional phase space. For no coupling a completely desynchronized motion was observed where the representative points associated with the oscillators spread over the whole phase space region allowed for an individual attractor. For the maximal coupling allowed, a synchronized state was observed with all the phase space points condensed in a single small ball which moved as a unity through phase space. The transition from the desynchronized state to the synchronized state was reported to be complex, and characterized by an intermediate state in which the set of 64 oscillators was divided in two stable subsets with the oscillators in each subset evolving in mutual synchrony and desynchronized from the oscillators of the other subset. Multistability was associated with this spatially nonhomogeneous state in the sense that the individual oscillators in each subset, as well as the size of the subsets were found to be different for different initial conditions; i.e. for different realizations of the experiment.

Moreover, for certain values of the coupling strength, this nonhomogeneous state was found to be unstable in the sense that the clusters change in composition and size as time evolves.

Chapter 7

Mutually Coupled Non-Identical Oscillators

When two different chaotic oscillators are mutually coupled, different types of synchronization phenomena may develop as the coupling strength is increased: amplitude envelope synchronization, phase synchronization, and lag, identical or generalized, synchronization can be observed. These forms of synchronization will be studied in this chapter, and relevant numerical and experimental work made on several systems, including chemical reactions, optical devices, and electric circuits will be discussed. Finally, a brief account of synchronization phenomena on networks of non-identical oscillators will be presented.

7.1 Degrees of chaotic synchronization

In most of this chapter, two parametrically different, and structurally equal chaotic oscillators labeled 1 and 2 will be considered. These are autonomous non linear dynamical systems whose dynamical states are given by the vectors $\mathbf{x}_1 \in \mathbb{R}^d$ and $\mathbf{x}_2 \in \mathbb{R}^d$, each of d scalar variables. Their dynamical evolutions are governed by the vectors fields $\mathbf{F}_1(\mathbf{x}_1)$ and $\mathbf{F}_2(\mathbf{x}_2)$ that have the same structure but parameters different enough as to allow us the consideration that the systems have different individual dynamics. The two systems are assumed to interact by a diffusive coupling so that their dynamics is given by the set of $2d$ ordinary nonlinear differential equations

$$\frac{d\mathbf{x}_1}{dt} = \mathbf{F}_1(\mathbf{x}_1) + \mathbf{C} \cdot (\mathbf{x}_2 - \mathbf{x}_1), \tag{7.1}$$

$$\frac{d\mathbf{x}_2}{dt} = \mathbf{F}_2(\mathbf{x}_2) + \mathbf{C} \cdot (\mathbf{x}_1 - \mathbf{x}_2), \tag{7.2}$$

with $\mathbf{C}$ a square matrix of dimension d whose constant elements, $C_{\alpha,\beta}$, measure the strength of the mutual coupling. This definition can be straightforwardly generalized to study chaotic oscillators that are structurally different, just by allowing the dimensions of the vector states to be different.

7.1.1 *Amplitude envelope synchronization*

Amplitude envelope synchronization is a form of synchronization that has been observed between two mutually coupled parametrically different oscillators [González-Miranda (2002a)]. It may appear alone, or combined with other forms of synchronization. In its pure form it develops at very low coupling strength, and it is quite a subtle form of synchronization, which does not introduce appreciable correlations between the amplitudes, or between the phases of the variables of the oscillators. The results correlated, in this case, are only the envelopes of equivalent variables of the two oscillators. This means that the extrema that each system variable can reach in a given time interval are correlated.

Given the time evolution of a chaotic dynamical variable, $s(t)$, its upper envelope is defined as the smoothest curve $\Sigma(t)$ that passes through all the maxima of $s(t)$. To obtain an estimate of the upper envelope from a finite sample of $s(t)$, limited to a time interval $t \in [t_0, t_f]$, all the times of occurrence of relative maxima of $s(t)$, $\tau^{(i)} \in [t_0, t_f]$, and the values of these maxima, $S^{(i)} = s\left(\tau^{(i)}\right)$, for $i = 1, 2, ..., N$ have to be recorded. The result is a discrete function $S(\tau)$ with values, $S^{(i)}$, defined for $\tau = \tau^{(i)}$ which all pertain to the envelope. If the time interval $[t_0, t_f]$ is large enough $S(\tau)$ provides a clear idea of how the upper envelope $\Sigma(t)$ is. In the same way the lower envelope can be defined from the minima of $s(t)$.

Amplitude envelope synchronization is the development of a correlation between the envelopes, $\Sigma_1(t)$ and $\Sigma_2(t)$, of the variables, $s_1(t) \in \mathbf{x}_1(t)$ and $s_2(t) \in \mathbf{x}_2(t)$, of the two coupled oscillators given by Eqs. 7.1–7.2. This correlation can be measured from the envelope estimates $S_1(\tau_1)$ and $S_2(\tau_2)$ through some appropriate correlation coefficient between two time series. To perform this calculation it is necessary to take into account that the envelopes estimates are recorded at series of time points that are different: $\tau_1^{(i)}$ and $\tau_2^{(j)}$, with $i = 1, 2, ..., N_1$ and $j = 1, 2, ..., N_2$. This can be made [González-Miranda (2002a)] by means of the use of surrogate series of data points which are defined by interpolation as follows: given $S_1(\tau_1)$ and $S_2(\tau_2)$, the surrogate of $S_1(\tau_1)$ relative to $S_2(\tau_2)$ is defined as the ordered set of points

$$\widetilde{S}_1^{(k)} = \frac{S_1^{(i)} \cdot \left(\tau_1^{(i+1)} - \tau_2^{(k)}\right) + S_1^{(i+1)} \cdot \left(\tau_2^{(k)} - \tau_1^{(i)}\right)}{\left(\tau_1^{(i+1)} - \tau_1^{(i)}\right)}, \tag{7.3}$$

with i given by the condition $\tau_1(i) \leq \tau_2(k) < \tau_1(i+1)$ and $k \in \mathcal{A} = \left\{j \,\middle|\, j = 1, 2, ..., N_2 \text{ and } \tau_1^{(1)} \leq \tau_2^{(j)} < \tau_1^{(N_1)}\right\}$. A correlation coefficient between $\widetilde{S}_1(\tau_2)$ and $S_2(\tau_2)$ is the quantity that is then computed for the purpose of detection of amplitude envelope synchronization. In particular, the linear correlation coefficient, given by [Press et al. (1992)]

$$r = \frac{\sum_{k\in\mathcal{A}} \left(\widetilde{S}_1^{(k)} - \left\langle \widetilde{S}_1 \right\rangle\right) \left(S_2^{(k)} - \langle S_2 \rangle\right)}{\sqrt{\sum_{k\in\mathcal{A}} \left(\widetilde{S}_1^{(k)} - \left\langle \widetilde{S}_1 \right\rangle\right)^2} \sqrt{\sum_{k\in\mathcal{A}} \left(S_2^{(k)} - \langle S_2 \rangle\right)^2}}, \tag{7.4}$$

where the angular brackets denote a time average, will be used here. This is a simple and well known coefficient, although not the only one which can provide an appropriate measure of the correlation [González-Miranda (2002a)]. The surrogate of $S_2(\tau)$ relative to $S_1(\tau)$, $\widetilde{S}_2^{(j)}$, can be defined in the same way, and should provide the same final result for the measure of the correlation between the signals.

A specific example of amplitude envelope synchronization will now be presented by means of the case of two diffusively coupled Van der Pol–Duffing oscillators [King and Gaito (1992); Gomes and King (1992)], labeled 1 and 2, whose equations of motion, written in dimensionless form, are:

$$dx_{1,2}/dt = -100\left(x_{1,2}^3 - 0.35x_{1,2} - y_{1,2}\right), \tag{7.5}$$

$$dy_{1,2}/dt = x_{1,2} - y_{1,2} - z_{1,2} + C\left(y_{2,1} - y_{1,2}\right), \tag{7.6}$$

$$dz_{1,2}/dt = \beta_{1,2}y_{1,2}, \tag{7.7}$$

with β_1, and β_2 parameters of the system which, in this case, have been fixed in $\beta_1 = 610$ and $\beta_2 = 680$. The coupling only affects the second equation of each oscillator, and its strength is measured by the constant C.

The development of amplitude envelope synchronization is presented in Fig. 7.1 by means of estimates of the envelopes of the system variables $z_1(t)$ and $z_2(t)$, $Z_1(\tau_1)$ and $Z_2(\tau_2)$, plotted for increasing values of the coupling strength. When the two oscillators evolve independently [Fig. 7.1(a)] the envelopes result in nearly random series of points with no apparent correlation between them. When the coupling is switched on, and increased, the

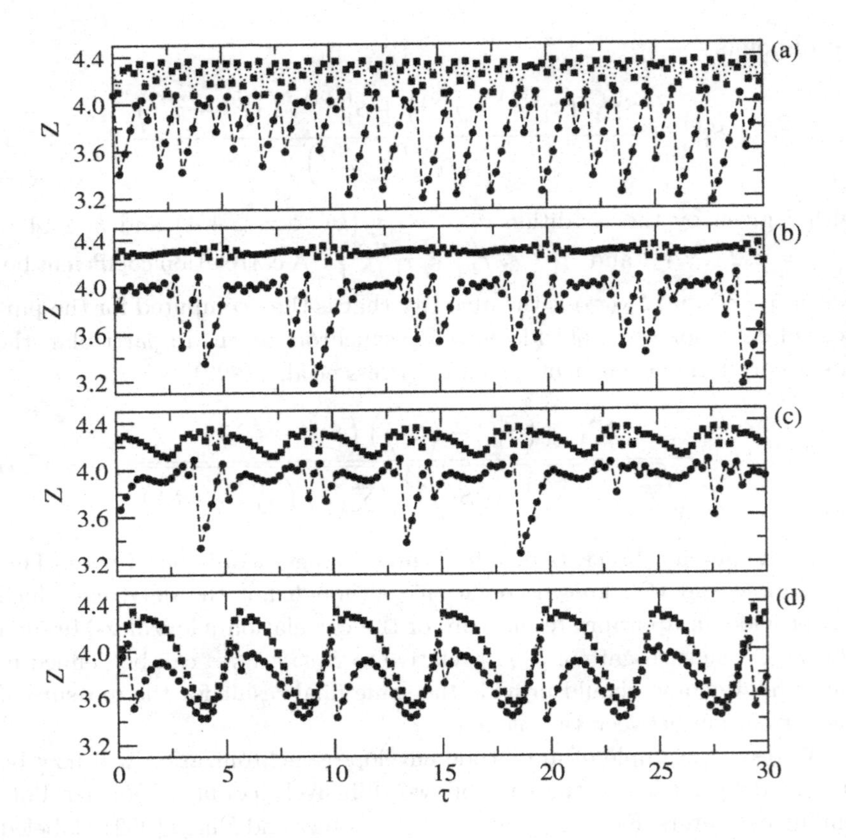

Fig. 7.1 Plots of the envelopes of $z_1(t)$ and $z_2(t)$ for increasing values of the coupling strength: (a) $C = 0$, (b) $C = 0.13$, (c) $C = 0.26$, and (d) $C = 0.39$. The circles refer to oscillator 1, and the squares to oscillator 2.

two envelopes develop a synchronized oscillation of the same period, which is disrupted by bursts that momentarily destroy the synchronization. This overall correlation is apparent, although very week, for $C = 0.13$ [Fig. 7.1(b)], results more unnoticeable for $C = 0.26$ [Fig. 7.1(c)], and is well developed for $C = 0.39$ [Fig. 7.1(d)], resulting in a synchronized modulation of the amplitudes of the variables $z_1(t)$ and $z_2(t)$.

More insight on amplitude envelope synchronization can be obtained from Fig. 7.2 which presents the dependence of relevant parameters of the dynamics of Eqs. 7.5–7.7 on the coupling strength. The absolute value of the linear correlation coefficient between the envelopes, $Z_1(\tau_1)$ and $Z_2(\tau_2)$, presented in Fig. 7.2(a) shows an overall increase of the correlation with the

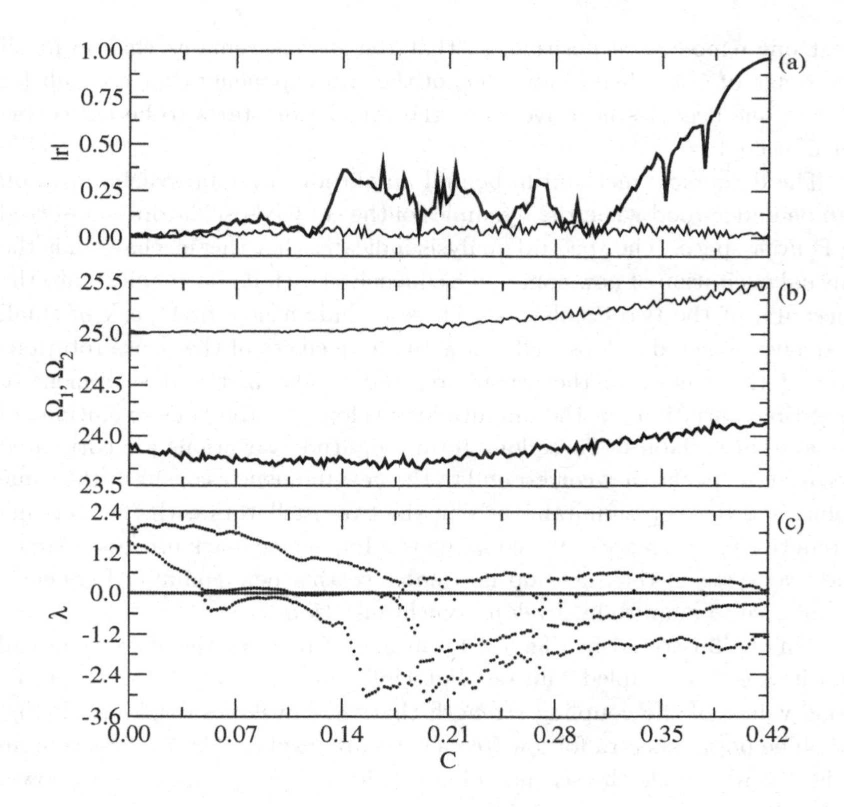

Fig. 7.2 Development of amplitude envelope synchronization with the coupling strength: (a) the linear correlation coefficient between $z_1(t)$ and $z_2(t)$ (thin line), and between $Z_1(\tau)$ and $Z_2(\tau)$ (thick line), (b) the frequencies associated to the phase of oscillator 1 (thick line) and of oscillator 2 (thin line), and (c) the four largest Lyapunov exponents of the six dimensional system that results from the coupling of the two oscillators.

coupling strength, which is in contrast with the same correlation coefficient computed between the signals, $z_1(t)$ and $z_2(t)$, which remains null for all values of C. In Fig. 7.2(b) the average frequencies for the phases of each oscillator, defined in Chapter 3, appear. They have been obtained from the return times of $z_1(t)$ and $z_2(t)$ through Eq. 3.11. There it is shown that the two frequencies remain almost constant in the coupling strength range that is observed. All this illustrates that amplitude envelope synchronization concerns the envelope of the amplitude of the oscillation rather than the oscillation itself. The four largest Lyapunov exponents of the six-dimensional system, given by Eqs. 7.5–7.7 and presented in Fig. 7.2(c), show that at least one exponent is positive, so that the system remains chaotic in all the range of C studied. Moreover, of the two exponents that are null for $C = 0$, one becomes negative when the correlation starts to be noticeable for $C \approx 0.04$.

The dynamical mechanism behind amplitude envelope synchronization can be understood when the dynamics of the coupled oscillators is observed in Fourier space. The spectral analysis indicates that this mechanism is the mutual excitation of new common frequencies (with its harmonics) into the spectrum of the two oscillators. These include a new first peak of small frequency which develops well below the frequencies of the phase rotations around the centers of the attractors, and results in the development of long time variations in the amplitude envelopes of the time evolutions of the system variables. These long term amplitude variations are correlated because the peaks that correspond to the new frequencies occur at the same value, and develop simultaneously in the two oscillators with the coupling strength. The increase of the coupling results in a feedback process between the two systems that develop and enhance this new common frequency, giving rise to amplitude envelope synchronization.

This is illustrated in Fig. 7.3 by means of plots of the power spectral densities of the coupled Van der Pol–Duffing oscillators computed at the same values of the coupling strength that the envelopes displayed in Fig. 7.1. The power spectra for low frequencies are displayed in the first column (Fig. 7.3(a)), while the second column (Fig. 7.3(b)) corresponds to power spectra for frequencies around the main peaks, which define the dominant frequencies of the chaotic oscillators. For $C = 0$, i.e. for the uncoupled systems, $P(\omega)$ shows a low and flat spectra at low frequencies, and two well defined main peaks with different frequencies at $\omega_1 \approx 23.9$, and $\omega_2 \approx 25.0$. When the coupling strength is switched on ($C = 0.13$), there appear new peaks in each spectrum at a small frequency, whose value is close to the difference of frequencies between the main peaks. Further increase in coupling strength ($C = 0.26, C = 0.39$) results in an enhancement of these new peaks and its harmonics, which is the cause of the development of synchronized modulations of the signals. Although, there is an additional mutual induction of peaks that correspond to the dominant frequencies, which develops further with the coupling strength, the dominant frequencies of each spectra continue to be the same. Therefore, the overall oscillating nature of the oscillations, regarding phase and amplitude, are still essentially the same as they were before the coupling.

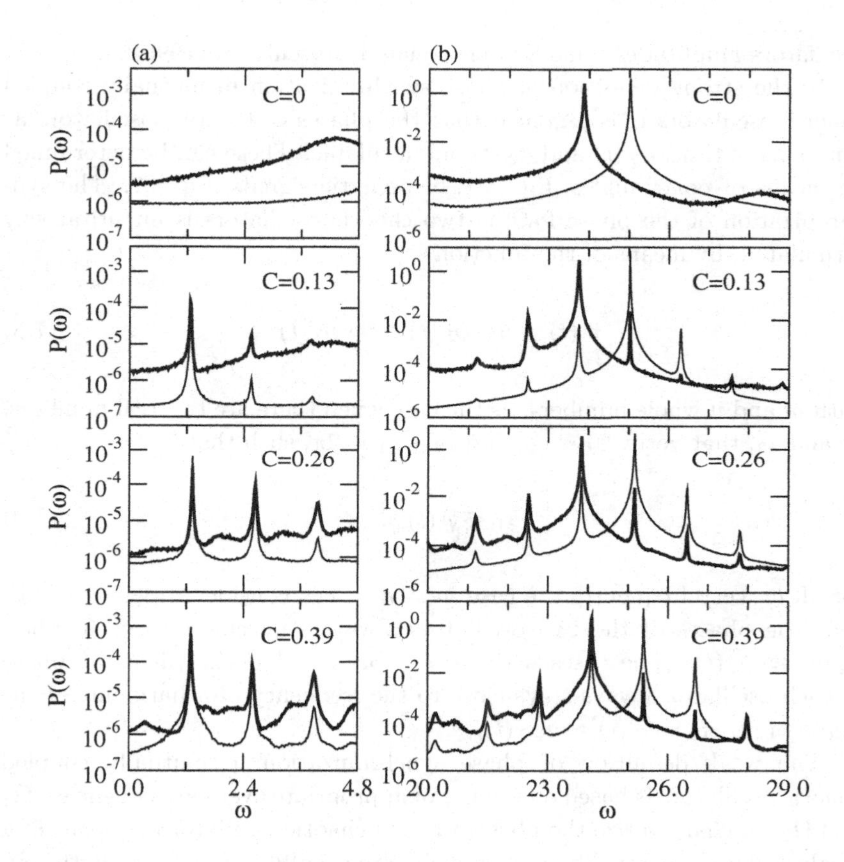

Fig. 7.3 Spectral analysis of the development of amplitude envelope synchronization. The power spectral density of oscillator 1 (thick line) and oscillator 2 (thin line) in the region (a) of small frequencies, and (b) of the dominant frequencies, both for the four values of the coupling strength, C, indicated in the figure.

7.1.2 *Phase synchronization*

The phenomenon of phase synchronization to an external periodic force, studied in Chapter 3, has its counterpart when non-identical mutually coupled chaotic oscillators are considered. The same definitions of the phase, and the same quantities and techniques used to observe and characterize the phase of a chaotic oscillator apply to each of the two oscillators that are coupled. The synchronization of the phase in this case means that, as an effect of the coupling, the phase dynamics of each oscillator becomes modified to evolve in pace with the phase of the other oscillator, while the oscillators amplitudes remain noncorrelated [Rosenblum (1996)].

In the strong definition of phase synchronization in mutually coupled chaotic oscillators it is assumed that the phases of the two oscillators as functions of time, $\phi_1(t)$ and $\phi_2(t)$, are available. These can be determined by means of polar angles, Eq. 3.1, or analytic signals, Eq. 3.7. The synchronization of the phase for the two chaotic oscillators is quantitatively formulated, by means of the function

$$\psi(t) = m \cdot \phi_1(t) - n \cdot \phi_2(t), \tag{7.8}$$

with m and n whole numbers, as the case when there are two real numbers, ε_1 and ε_2, that verify $\varepsilon_1 < \varepsilon_2$, and $\varepsilon_2 - \varepsilon_1 < 2\pi$, such that

$$\varepsilon_1 < \psi(t) < \varepsilon_2 \tag{7.9}$$

for all t. Very frequently the case $m = n = 1$ is considered, which means that the phases of the two oscillators always stay closer to each other: $\phi_1(t) \approx \phi_2(t)$. The cases with $m \neq 1$ or $n \neq 1$ occur when the phase of each oscillator always stays close to the frequency of a harmonic of the phase of other, $m\phi_1(t) \approx n\phi_2(t)$.

The weak definition of phase synchronization in mutually coupled chaotic oscillators is based on the use of appropriate average frequencies, Ω_1 and Ω_2, to characterize the phases of each chaotic oscillator by means of a single scalar quantity. These frequencies are usually computed from the average periods (Eq. 3.11) obtained as the average times between crossings of a Poincaré sections, or between the returns of a variable. Stroboscopic plots (Fig. 3.3) can also be used to determine these frequencies. The condition for phase synchronization is then

$$m \cdot \Omega_1 - n \cdot \Omega_2 = 0 \tag{7.10}$$

with the same interpretations as in the above paragraph for the cases $m = n = 1$, and $m \neq 1$ or $n \neq 1$.

These ideas will now be illustrated by means of an example based on the Rössler model [Rössler (1976)] which, as seen in Chapter 3, for appropriate parameter values, presents a chaotic attractor whose dynamics follow a proper rotation with a phase that can be characterized by means of any of the above procedures. The particular form of Eqs. 7.1–7.2 used is

$$dx_{1,2}/dt = -\left(\omega_{1,2}y_{1,2} + z_{1,2}\right), \tag{7.11}$$
$$dy_{1,2}/dt = \omega_{1,2}x_{1,2} + ay_{1,2} + C\left(y_{2,1} - y_{1,2}\right), \tag{7.12}$$
$$dz_{1,2}/dt = b + z_{1,2}(x_{1,2} - c), \tag{7.13}$$

where the original Rössler model (Eqs. 2.23–2.25) has been modified by the introduction of the new parameters ω_1 and ω_2 which control the average frequency of rotation of each oscillator. They can be written as $\omega_1 = \Omega_0 + \Delta$ and $\omega_2 = \Omega_0 - \Delta$, with Ω_0 a frequency of reference, usually close to 1, and Δ half of the difference of the frequency of the oscillators. This modification of the Rössler model is aimed to have the possibility of working with coupled oscillators which are very alike between them, being their main difference the rate of the phase dynamics. The coupling, of strength C, acts on the second equation of each oscillator, and the parameter values used in the example are $a = 0.16$, $b = 0.1$, $c = 8.5$, $\Omega_0 = 1$, and $\Delta = 0.02$.

The strong definition of phase synchronization can be studied in the $x - y$ plane, where the attractor follows a simple counterclockwise rotation around the origin which makes the dynamics of the phase simple, and the analysis quite intuitive. The results presented in Fig. 7.4 are for polar angles measured in this plane, $\phi_{1,2}(t) = \arctan\left[y_{1,2}(t)/x_{1,2}(t)\right]$, which provide an appropriate measure of the phase. The dynamics of $\psi(t) = \phi_1(t) - \phi_2(t)$, for several values of the coupling strength is displayed in Fig. 7.4(a), this shows that for uncoupled oscillators the phases evolve independently, and its difference growths steadily. When the coupling is switched on the rate of the divergence between $\phi_1(t)$ and $\phi_2(t)$ decreases, and $\psi(t)$ is composed of time intervals of steady behavior that are disrupted by jumps of 2π. Two regimes can be observed [Lee et al. (1998)]: for weak coupling the sequence of phase slips is approximately periodic ($C = 0.030$), and when the transition to phase synchronization is closely approached the intermittence of phase jumps becomes irregular ($C = 0.038$). Finally a state of strong phase synchronization [Rosenblum (1996)], with a $\psi(t)$ steady and bounded, has been achieved ($C = 0.045$). This state of phase synchronization is further illustrated in Fig. 7.4(b) which shows that the distribution function of the values of $\psi(t)$ in this state is limited to a narrow region well within the interval $(-\pi, \pi)$. Fig. 7.4(c) presents the parametric plot of x_2 versus x_1 in this state of phase synchronization: the spread of points shows that it is far from being a case of identical synchronization, i.e. there is correlation between the phases, but not between the amplitudes. This is further illus-

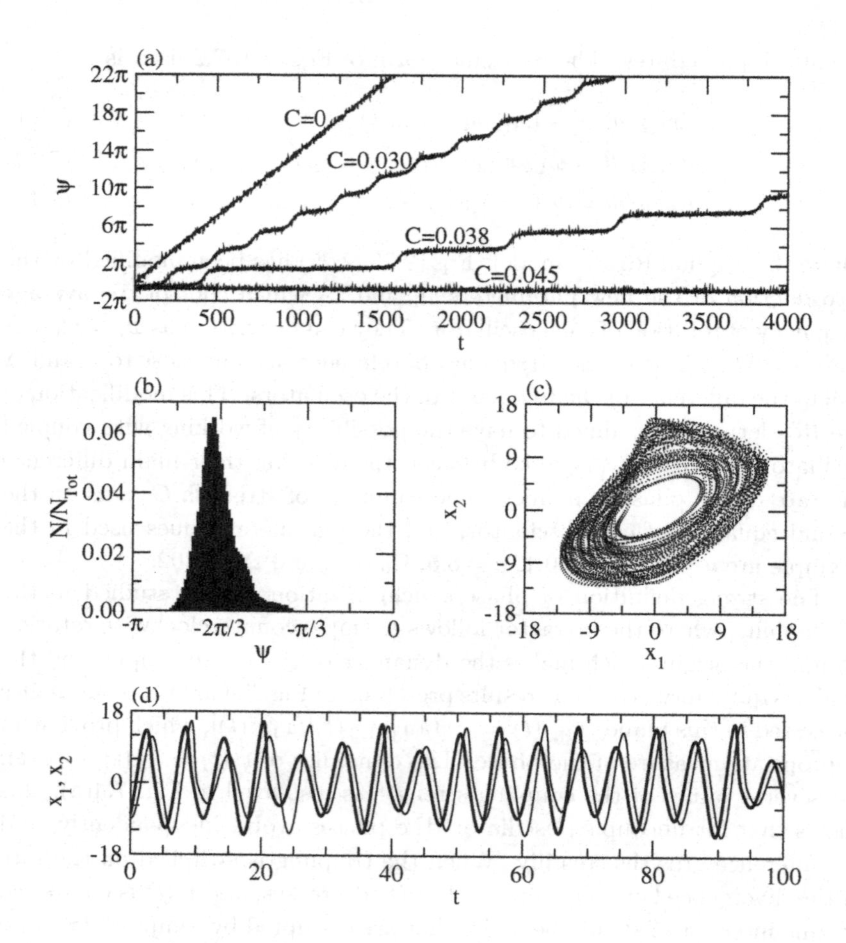

Fig. 7.4 Phase synchronization in the modified Rössler model according to the strong definition. (a) Dynamics of the phase difference, $\psi(t)$, for four values of the coupling strength, C, indicated in the plot. Characterization of a state of phase synchronization ($C = 0.045$) by: (b) the histogram of the values reached by $\psi(t)$, (c) a test for identical synchronization made by the parametric plot of $x_2(x_1)$, and (d) visualization of the phase synchronized state by time series of $x_1(t)$ (thick line) and $x_2(t)$ (thin line).

trated in Fig. 7.4(d) which presents the time series for the variables $x_1(t)$ and $x_2(t)$ in the state of phase synchronization: the time of occurrence of the peaks of each signal are correlated, although there is a certain lag between consecutive maxima; however, the heights of the maxima do not appear correlated.

The weak definition of phase synchronization applied to this system

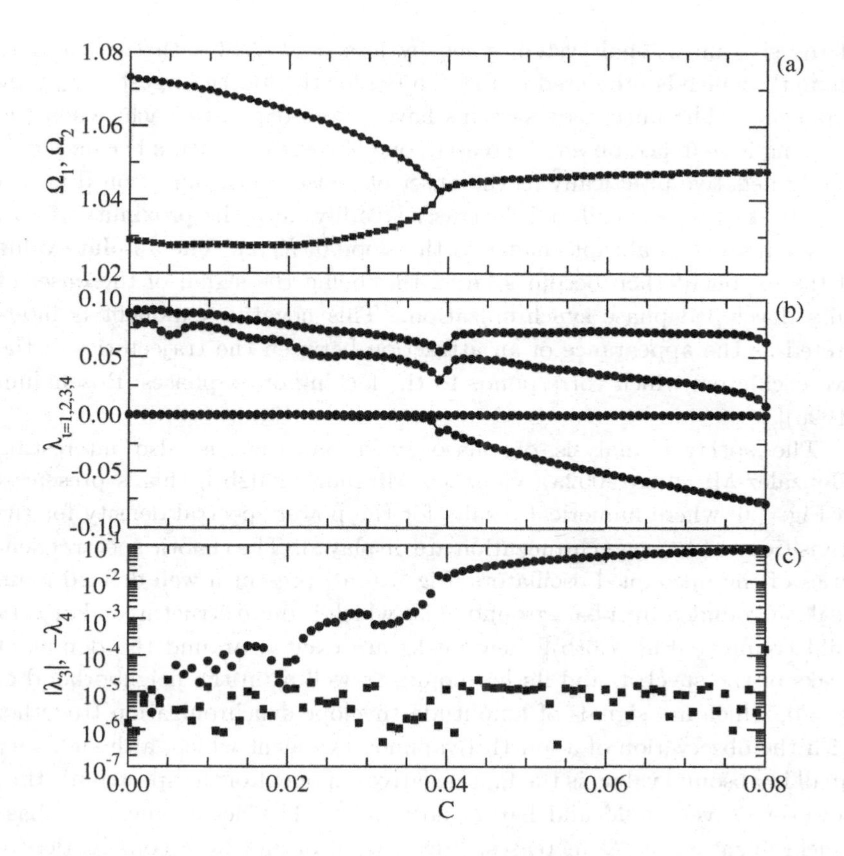

Fig. 7.5 Phase synchronization in the modified Rössler model according to the weak definition: (a) dependence on the coupling strength of the frequencies of the phases of each oscillator obtained from the return times of $z_1(t)$ (squares) and $z_2(t)$ (circles). Lyapunov stability analysis of phase synchronization as a function of the coupling strength: (b) the fourth largest Lyapunov exponents, and (c) a detailed view of the behavior of the third (squares), and fourth (circles) exponents.

is illustrated in Fig. 7.5. The phase has been characterized by means of the frequencies, $\Omega_1 = 2\pi/T_1$ and $\Omega_2 = 2\pi/T_2$, computed from the average values, T_1 and T_2, of the time intervals between maxima of $z_1(t)$ and $z_2(t)$. Their dependence on the coupling strength, presented in Fig. 7.5 (a), shows a smooth evolution of the values of Ω_1 and Ω_2 to match each other at certain critical value $C_P \approx 0.039$. For values of C greater than C_P the frequencies remain locked. Additional illustration of this phenomenon of phase synchronization is provided by the change of the Lyapunov spectrum of the six-dimensional system given by Eqs. 7.11–7.13 with the coupling strength, which is presented in Fig. 7.5 (b) for the fourth largest Lyapunov exponents. The uncoupled systems have a zero exponent each; when the coupling is switched on and increased, one of these exponents becomes definitely negative practically at the onset of phase synchronization [Fig. 7.5 (c)]. It stays very small and decreases steadily until the proximity of C_P, where there is an abrupt change in the slope of $\lambda_4(t)$. The absolute value of the exponent then becomes large, this being the signal of the onset of fully developed phase synchronization. This negative exponent is interpreted as the appearance of an attraction between the trajectories of the two oscillators which corresponds to the locking of its phases [Rosenblum (1996)].

The spectral analysis of phase synchronization is also interesting [González-Miranda (2002a); González-Miranda (2002b)]; this is presented in Fig. 7.6, where numerical results for the power spectral density for the transition to phase synchronization are displayed. The chaotic spectral densities of the uncoupled oscillators [Fig. 7.6(a)] present a well defined main peak surrounded by a background noise which is quite structured. For very mild coupling [Fig. 7.6(b)], new peaks are excited around the dominant peaks of the spectra, and its harmonics as well as in the neighborhood of $\omega = 0$, which are signals of amplitude envelope synchronization (together with the observation of a fourth Lyapunov exponent which, although very small in absolute value, is the first negative exponent of the spectrum); this, however, is very mild and hardly noticeable. The development of phase synchronization, as C increases, [Fig. 7.6(c)] occurs by a combination of the broadening of the peaks with the increase in the number of induced peaks, which results in the shift of the main peak of each spectrum (and its harmonics) to match the corresponding peak of the other [Fig. 7.6(d)].

This scenario bears great resemblance with that shown in Fig. 3.6 for periodic driving, the main difference being that, in the present case, there is a feed-back between the two systems which lets them lock to each other. Therefore, the spectral analysis of the dynamics of the two scenarios reveals the same underlaying mechanism. Moreover it shows that the locking of the phase in not exactly a threshold phenomenon, but a process of mutual locking between oscillatory motions in the two systems, that develops steadily until a state of fully developed synchronization settles down when the oscillations correspondent to the main frequencies result finally locked. This progressive development of phase synchronization is an aid to understanding the behavior of $\lambda_4(t)$ mentioned in the previous paragraphs.

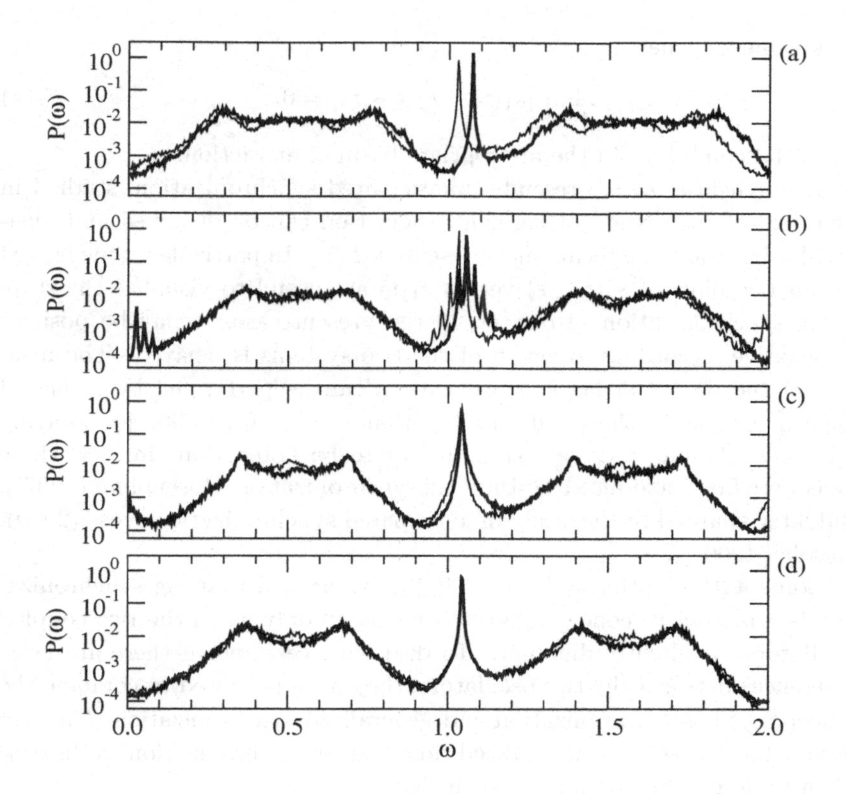

Fig. 7.6 Power spectral density analysis of two mutually coupled modified Rössler oscillators for four values of the coupling strength: (a) $C = 0$, (b) $C = 0.030$, (c) $C = 0.038$, and (d) $C = 0.045$. The thick line is for oscillator 1 ($\omega_1 = 0.98$), and the thin line of oscillator 2 ($\omega_2 = 1.02$).

7.1.3 *Lag synchronization*

Beyond phase synchronization there is a state of enhanced synchronism know as lag synchronization [Rosenblum (1997)]. In the phase synchronized state, the maxima (or minima) of the system dynamical variables, $s_1(t)$ and $s_2(t)$, besides being noncorrelated in amplitude, do not coincide in time [Fig. 7.4(d)]: the signal from the system with a lower frequency has some delay with respect to the system which has the higher frequency. Further increase of the coupling may lead to a correlation between the amplitudes of the variables that describe the system, although still being a shift between them, so that a relation of the type $s_2(t - \tau) \approx s_1(t)$, with τ a time lag between the systems, develops. Chaotic lag synchronization has been achieved when

$$\lim_{t\to\infty} |s_1(t) - s_2(t-\tau)| = 0, \tag{7.14}$$

for initial conditions in the appropriate basin of attraction.

Lag synchronization resembles anticipated synchronization, studied in Chapter 4, and a state of lag synchronization can be observed and measured using the same techniques presented there. In particular, time lagged parametric plots of $s_2(t-\tau)$ versus $s_1(t)$ are useful to visualize the state of lag synchronization. However, in the present case, τ may be positive or negative, according to which of the two systems is delayed. The measure of the degree of lag synchronization can be performed by means of the calculation of the similarity function, $S(T)$, Eq. 4.30, just setting $x(t) = s_1(t)$ and $x'(t) = s_2(t)$. It has to be noted that, in fact, these tools were first introduced to study lag synchronization [Rosenblum (1997)], and later adapted to the study of anticipated synchronization [Voss (2000); Masoller (2001)].

Some authors [Zhu and Lai (2001)] have argued that lag synchronization is a restrictive concept that will be useful only when the two coupled oscillators are slightly different. To deal with cases when there are large differences between the two oscillators, they have proposed to combine the concepts of lag synchronization and generalized synchronization (Chapter 4) to define the state of generalized time-lagged synchronization as the case when there is a functional $\phi[\cdot]$ such that

$$\lim_{t\to\infty} |s_1(t) - \phi[s_2(t-\tau)]| = 0, \tag{7.15}$$

is fulfilled. Of course, this condition has Eq. 7.14 as the particular case when $\phi(s)$ is the identity. Consequently, a generalized similarity function, $S_G(T)$, has been introduced to measure generalized time lagged synchronization. Given the signals $s_1(t)$ and $s_2(t)$, and using T to denote the time shift between them this is defined by means of

$$S_G^2(T) = \frac{\left\langle \{s_1(t) - \phi[s_2(t-T)]\}^2 \right\rangle}{\sqrt{\left\langle [s_1(t)]^2 \right\rangle \left\langle \phi[s_2(t)]^2 \right\rangle}} \tag{7.16}$$

where $\langle\cdot\rangle$ denotes a time average, and $s_1(t)$ and $s_2(t)$ are assumed to have zero mean. The similarity function takes values of the order of 1 when $s_1(t)$ and $\phi[s_2(t-T)]$ are not correlated, and becomes null when $s_1(t)$ and $\phi[s_2(t-T)]$ are identical. Synchronization is then detected by finding the values of $T = T_m$ where $S_G(T)$ presents a minimum, and the degree of synchronization is measured by the closeness of $S_{\min} = S_G(T_m)$ to zero. The main difficulty with the practical application of Eq. 7.16 is that, here, besides $\tau = T_m$, the functional for $\phi[\cdot]$ also has to be determined. Nonetheless, this expression is useful for theoretical analysis, and may be used in practice by a method of successive approximations starting with $\phi[\cdot]$ equal to the identity [Zhu and Lai (2001)].

An example of lag synchronization is now presented. This continues the above example of phase synchronization on two modified Rössler oscillators mutually coupled (Eq. 7.11–7.13), by going to larger coupling strength. The observation of a particular lag synchronized state is presented in Fig. 7.7. The similarity function (Eq. 4.30) computed for $x_1(t)$ and $x_2(t)$ for $C = 0.110$ is presented in Fig. 7.7(a). For $\tau \approx -0.36$ there is an acute minimum where $S(\tau) \approx 0$: this is an indication of a synchronized state in which system 2 reproduces the dynamics of systems 1 with a time delay of 0.36; i.e. a case of lag synchronization. For the sake of comparison, in the same figure the results for the similarity function in the phase synchronized state ($C = 0.045$), the desynchronized state ($C = 0.030$), and for the uncoupled systems appear. In this last case, $S(t)$ is almost flat, while in the two other cases minima, which are far of being null indicate that in these cases there is not lag synchronization. This is illustrated in Figs. 7.7(b, c) by means of parametric plots of the time lagged variable $x_2(t-\tau)$ versus $x_1(t)$ with the value of the time lag given by the minima in Fig. 7.7(a) for each case: cloudy distributions of points appear for the desynchronized and the phase synchronized cases, indicating that these signals are practically uncorrelated. However, in the state of lag synchronization [Fig. 7.7(d)] a 45° slope segment of a straight line tells that $x_1(t) = x_2(t-\tau)$. This is clearly appreciated in the time series presented in Fig. 7.7(e) where it is seen that $x_1(t)$ is a copy of $x_2(t)$ delayed a small amount τ.

The development of lag synchronization, in this particular example, is monitored by means of the variation of several quantities with the coupling strength presented in Fig. 7.8. The dependence of the value of the minimum of the similarity function, $S(\tau)$, with the coupling strength, C, shown in Fig. 7.8(a), indicates that this quantity decreases with the coupling strength, and abruptly becomes null at certain coupling $C_L \approx 0.098$ which indicates the effective start of the chaotic time lagged synchronized state. The absolute value of the time lag between the signals [Fig. 7.8(b)] decreases monotonously in absolute value, and is quite small in the lag synchronized state, which evolves asymptotically towards a state of iden-

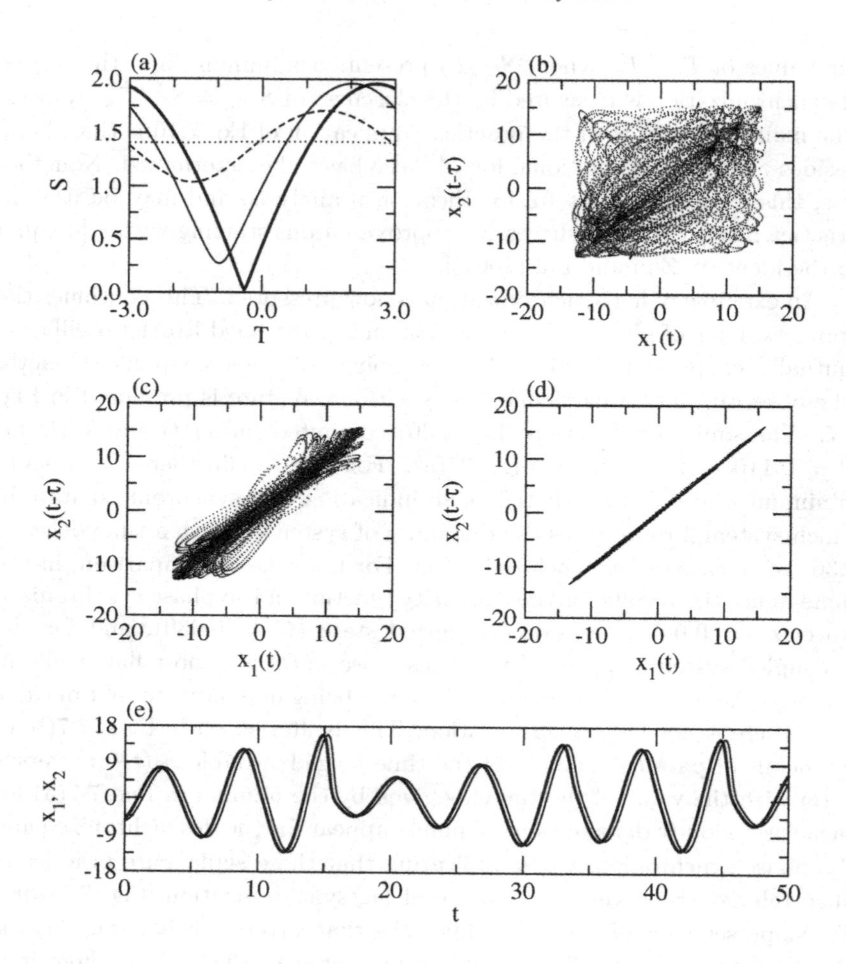

Fig. 7.7 (a) The similarity function for two Rössler oscillators mutually coupled for four different coupling strengths: $C = 0$ (dotted line), $C = 0.030$ (dashed line), $C = 0.045$ (thin line), and $C = 0.110$ (thick line). Parametric plots for the appropriate delayed coordinates at different synchronization regimes: (b) no synchronization ($C = 0.030$), (c) phase synchronization ($C = 0.045$), and (d) lag synchronization ($C = 0.110$). Time series for the system variables $x_1(t)$ (thick line), and $x_2(t)$ (thin line) in a time lagged synchronized state.

tical synchronization as the coupling strength increases. The frequency difference between the two oscillators continues to be null in the phase synchronized state [Fig. 7.8(c)], and the second Lyapunov exponent [Fig. 7.8(d)], which was positive in the chaotic phase synchronized state, now has

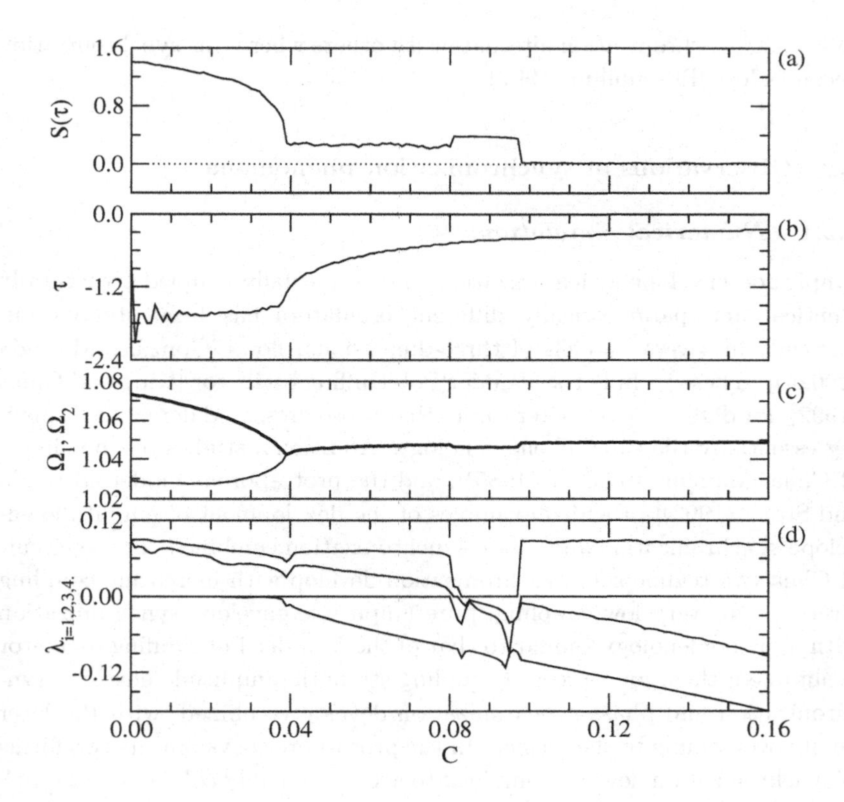

Fig. 7.8 Development of lag synchronization with increasing coupling strength monitored by: (a) the minimum value of the similarity function, and (b) its time of occurrence, (c) comparison of the frequencies of the two oscillators obtained from the return times, and (d) the four largest Lyapunov exponents.

become negative and is the third exponent. However, the crossing of the $\lambda = 0$ line occurs at $C'_L \approx 0.08$, where a window of phase locked periodic motion starts, and only when this window closes at C_L, the chaotic lag synchronized regime begins. It is to be noted that this transition through a periodic window is not the general case, but more an example that the kind of complexity studied in Chapter 6 can also appear in the present context. For other system parameter values, although the appearance of periodic windows is common, it is possible to find transitions from phase to lag synchronization where the system stays chaotic. In this case however, the crossing of the $\lambda = 0$ line still occurs at $C'_L < C_L$, because for C in the interval $[C'_L, C_L]$ a regime of intermittence develops where time intervals of lag synchronization alternate with others where this synchronization becomes lost [Rosenblum (1997)].

7.2 Observations of synchronization phenomena

7.2.1 *Numerical simulations*

Amplitude envelope synchronization of two mutually coupled structurally identical and parametrically different oscillators has been studied numerically in several models of three-dimensional flows [González-Miranda (2002a)]. These include the Van der Pol–Duffing oscillator [King and Gaito (1992)] for different parameter values that those presented here, but displaying essentially the same phenomenology. Additional studies on the circuit of Chua [Matsumoto et al. (1985)], and the proto-Lorenz model [Miranda and Stone (1993)] provided examples of the development of amplitude envelope synchronization and phase synchronization combined. In the circuit of Chua two regimens of synchronization develop with increasing coupling strength: for very low coupling, pure amplitude envelope synchronization with a phenomenology similar to that of the Van der Pol–Duffing oscillator is obtained; then, for increased coupling strength, amplitude envelope synchronization and phase synchronization develop combined, with the later finally overcoming to the former. In the proto-Lorenz system the two forms of synchronization develop combined to achieve a highly coherent state with both, phases and envelopes of the amplitudes, synchronized.

Numerical studies of phase and lag synchronization have been performed mainly on sets of two coupled modified Rössler models, being the overall rotational frequency, given by ω_1 and ω_2, the difference between the oscillators. Although these investigations [Rosenblum (1996); Rosenblum (1997)] were made tuning the Rössler model at other parameters then those reported here, and coupling through the x variable, instead of through the y, the phenomenology observed is essentially the same presented in this chapter: phase synchronization reached by intermittence and then lag synchronization. There is only one difference, this is the transition from phase to lag synchronization, which these authors observed to be through an intermittence route, instead of after a chaos suppression episode. Other authors [Lee et al. (1998)] have also demonstrated and characterized an intermittence route for the transition from the desynchronized state to phase synchronization.

Phase synchronization has also been studied between two parametrically different Lorenz models coupled through the x variable, and observed in the $u-z$ plane, being $u=\sqrt{x^2+y^2}$, which provides an example weak phase coherent attractor. An analysis in terms of the strong definition of phase synchronization [Lee et al. (1998)] provided a picture of the transition to phase synchronization, similar to that of the Rössler model: an intermittent route of phase slips that become less frequent as the synchronized state is approached. The analysis along the weak definition of phase synchronization [Chen et al. (2001)] also resulted in a picture of two approaching frequencies until they match each other at a critical value of the coupling strength. However, in this case the behavior of the Lyapunov spectra was different than in the most phase coherent Rössler model in that the critical point was signaled by a positive Lyapunov exponent becoming negative.

Besides those cases of phase synchronization in coherent, and weakly incoherent, chaotic oscillators, the synchronization of the phase of chaotic oscillators whose dynamics are not phase coherent has started to be studied recently. For this aim a special definition of the phase has been introduced [Chen et al. (2001)], based on the use of Eq. 3.1, but applied, not in the plane $x-y$ of coordinates of the oscillator, but in the plane $\dot{x}-\dot{y}$ of velocities. This has allowed the numerical observations of phase synchronization in the funnel attractor of the Rössler system [Chen et al. (2001)]. Further development based on this definition has been useful to identify three types of transition to phase synchronization, dependent on the degree of coherence of the oscillators that are coupled, which have been illustrated by means of the Rössler system working in regular and in funnel attractors [Osipov et al. (2003)].

Synchronization between two coupled chaotic structurally different systems has also received certain attention. The study of the dynamics of a Rössler chaotic oscillator [Rössler (1976)] and a Rössler hyperchaotic oscillator [Rössler (1979)] mutually coupled has provided an example of proper weak phase synchronization of chaos [Rosenblum (1996)] understood as the case when the difference between average frequencies, $\Omega_2-\Omega_1$, becomes null, while the phase difference $\psi(t)$ is not bounded by the condition given by Eq. 7.9. Another study [Boccaletti et al. (2000)] was focused on the dynamics of two coupled high dimensional systems with individual dynamics that occur in spaces that have different dimensions. For this aim the fact that delayed dynamical systems, such as the Mackey–Glass model [Mackey and Glass (1977)], have dynamics whose dimension can be controlled by changing the delay time was used. Starting with two Mackey–Glass oscillators whose dimensions were different and greater than 10, diffusive coupling was switched on, and as a function of the coupling strength a systematic decrease of the dimension of the compound system was observed. For weak coupling this resulted in a synchronized state that can be described as generalized phase synchronization, with no lag. For large coupling, chaos was finally destroyed and the resulting dynamics was that of two synchronized periodic oscillators.

7.2.2 *Laboratory experiments*

Phase synchronization between two coupled non identical oscillators has also been observed in the laboratory using chaotic oscillators of different natures. One observation using chemical oscillators was made using a standard electrochemical cell similar to that described when phase synchronization to an external periodic force was studied in Chapter 3. In the present case [Kiss and Hudson (2002)], the experimental set-up included two working electrodes embedded in an insulator that only lets free the tips. The chemical reactions that occur at the end of each electrode were the two chaotic oscillators studied. The electrodes were coupled to the reference electrode and to the counter electrode by means of a potentiostat and a common resistance which could be varied to control the strength of the coupling between them, and then between the chemical reactions at its ends. The currents of the electrodes were measured, and used as the experimental observables to obtain the chaotic attractors through embedding techniques, and the phases of the oscillators by means of the analytic signal method. The two independent chaotic attractors had the same structure, the main difference being the values of the frequency of rotation. For weak coupling, phase synchronization was identified by means of the stationary behavior of the phase difference, $\psi(t) = \phi_1(t) - \phi_2(t)$, and the independence of the amplitudes of the analytic signals (observed in parametric plots). For increased coupling strength, a state of identical or nearly identical synchronization characterized by phase locking and a development of a correlation between the amplitudes was observed.

A second observation was made in a thermo-optical system [Herrero et al. (2002)] made up an optical cavity limited by two parallel mirrors, and filled with three layers of transparent materials parallel to the mirrors and arranged in such a way that two consecutive layers had opposite thermo-optic coefficients. The mirrors closing the cavity have different properties: the rear mirror has high reflecting properties while the front mirror partially absorbs the incident power. In the experiments, this device is illuminated by a laser beam perpendicular to the layers, and the reflected power is measured as a function of time. The power absorbed in the front mirror combined with the nonlinearity inherent to the light interference process may give rise to chaotic oscillations in the reflected power. That means that locally, in the region illuminated by the laser beam there is a thermo-optical chaotic oscillator [Herrero et al. (1996)]. Two coupled chaotic oscillators of this kind were created by the use of two parallel laser beams acting on the same cavity at close distance to allow thermal conduction to establish an energetic link between the oscillators [Herrero et al. (2002)]. The oscillators were different because of minor local incontrollable differences inherent to the experimental set up, combined with induced differences achieved by independently tuning the two laser beams. The system was studied as a function of the total input power provided by the laser beams. The synchronization of the phase was monitored by means of the observation of time series of the reflected power. The phase difference, $\psi(t)$, obtained from the difference between the maxima of these time series, their power spectra, and the Lyapunov exponents obtained by means of embedding techniques were the elements used to detect phase synchronization. Two experiments, made for different temperatures, displayed characteristic phenomena of phase synchronization like those described in Subsection 7.1.2: a bound evolution of $\psi(t)$ achieved by an intermittence route, development of matching maxima in the power spectra, and a null exponent becoming negative before the onset of fully developed phase synchronization.

Experimental observations of lag synchronization have been performed, although certain difficulties are present. In one case, systems made of two mutually coupled electric circuits, each designed to realize the dynamics of a Rössler oscillator with different system parameters, were implemented [Taherion and Lai (1999)]. Lag synchronization was characterized by means of the similarity function, Eq. 4.30, and the difference $|x_2(t+\tau) - x_1(t)|$. This allowed the observation of lag synchronization in an interval of values of the coupling strength; however, synchronization had an intermittent nature with bursts of desynchronized behavior disturbing the state of lag synchronization. For larger coupling strength identical synchronization was reported. The intermittence, in this case, was interpreted [Taherion and Lai (1999)] as an effect of an unavoidable level of noise, intrinsic to the experimental realization, which prevented the observation of any kind of lag synchronization when the mismatch between the oscillators was smaller than the noise level. This result was substantiated with numerical simulations of noisy Rössler models mutually coupled. In another case, a set up based on the circuit of Chua allowed the observation of lag synchronization [Zhu and Lai (2001)]; however, a generalized similarity function [Eq. 7.16] had to be used, indicating that generalized time lagged synchronization was in fact observed.

7.3 Systems of $N > 2$ non-identical oscillators

In Chapter 6, systems of several or many coupled identical or nearly identical oscillators were studied. To deal with many cases commonly found in practice a generalization is needed to the case when the oscillators are not identical. In this case, N structurally identical and parametrically different chaotic flows, $d\mathbf{x}_i/dt = \mathbf{F}_i(\mathbf{x}_i)$ with $i = 1, 2, ..., N$, will be coupled together in an arrangement usually modelled by

$$\frac{d\mathbf{x}_i}{dt} = \mathbf{F}_i(\mathbf{x}_i) + C \cdot \sum_{j=1}^{N} c_{i,j} (\mathbf{x}_j - \mathbf{x}_i), \; i = 1, 2, ..., N \tag{7.17}$$

The expression for the coupling term in this equation is equal to the equivalent term in Eq. 6.18. The discussion presented in Section 6.2 on the different particular forms of realization of this coupling, and on boundary conditions still holds here. Moreover, in this case, the set of N oscillators can be realized in different forms because the system parameter values have to be chosen according to certain deterministic or stochastic rules which can differ from one system to another. So, when dealing with systems of non-identical oscillators, to have a well defined problem, besides the coupling matrix $c_{i,j}$ which describes the configuration of the interactions, a new element also has to be specified: the structure of the population of oscillators.

Phase synchronization is the phenomenon mostly studied in this case. It is known that two mutually coupled non-identical chaotic oscillators may achieve a state in which the phases are synchronized and the amplitudes uncorrelated; therefore, an analogous phenomenon of phase synchronization is expected on systems of $N > 2$ non-identical chaotic oscillators that are coupled. This synchronized state can be characterized by a frequency, Ω, common to all oscillators describing its phase dynamics, so that a condition for phase synchronization commonly used is

$$\Omega_1 = \Omega_2 = ... = \Omega_N = \Omega, \tag{7.18}$$

with these individual frequencies defined for each oscillator by any of the techniques presented in Chapter 3.

To present an example of phase synchronization in a network of chaotic oscillators, a system of $N = 18$ Rössler systems mutually coupled to nearest neighbors by diffusive coupling of the y variable, and subject to periodic boundary conditions will be considered. The Rössler oscillators are of the modified type considered in Eqs. 7.11–7.13 with the structure of the population of oscillators defined by means of the frequency parameters, $\omega_{i=1,2,...,N}$, which are chosen according to the rule

$$\omega_{i+1} = \begin{cases} \omega_i + \Delta\omega, & \text{for } i \leq 9 \\ \omega_i - \Delta\omega, & \text{for } i > 9 \end{cases} \tag{7.19}$$

with $\omega_1 = 1$, and $\Delta\omega = 0.0003$. The equations of motion are then

$$dx_i/dt = -\left(\omega_i y_i + z_i\right), \tag{7.20}$$

$$dy_i/dt = \omega_i x_i + a y_i + C\left(y_{i+1} + y_{i-1} - 2y_i\right), \tag{7.21}$$

$$dz_i/dt = b + z_i(x_i - c), \tag{7.22}$$

with $i = 1, 2, ..., N$, and the choice of Rössler system parameters $a = 0.16$, $b = 0.1$, and $c = 8.5$.

The process of phase synchronization is monitored by means of the average frequency of each oscillator computed from the return times (Eq. 3.11) of $x_i(t)$, T_i, as $\Omega_i = 2\pi/T_i$, for $i = 1, 2, ..., N$. The plot of the values of these frequencies in the uncoupled state ($C = 0$), displayed in Fig. 7.9(a), shows the shape of an inverted V induced by the spectra of the frequency parameters, $\omega_{i=1,2,...,N}$, given by Eq. 7.19. At very small values of the coupling strength a phenomenon of partial synchronization of the phase develops Fig. 7.9(b), in which the phase synchronization condition given by Eq. 7.18 separately develops in two clusters of oscillators, one of high frequencies, in the middle of the network, and the other of low frequencies, in the tips of the line, connected by the periodic boundary conditions. As the coupling strength increases the values of these two frequencies continuously tend to match each other. For larger coupling strength [Fig. 7.9(b)] the two clusters have merged to give a state of complete phase synchronization, in which all the oscillators have the same frequency; i.e. Eq. 7.18 is fulfilled by all oscillators. The value of this common frequency steadily decreases with increasing coupling strength.

The synchronization of the phases, however, does not imply the synchronization of the amplitudes. This is illustrated in Fig. 7.10 where the time series of the x-signal of the first three oscillators of the chain are presented for two values of the coupling strength. In Fig. 7.10(a) the coupling

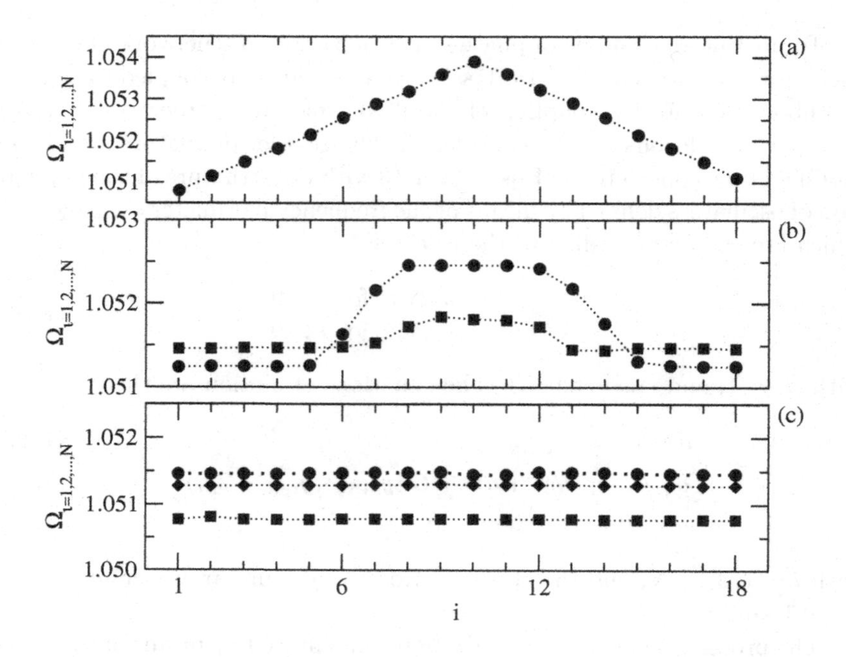

Fig. 7.9 Frequencies of the oscillators of a chain of $N = 18$ diffusively coupled Rössler oscillators for different values of the coupling strength, C. Each oscillator is indicated by a symbol, and identified by the number, i, of its position in the chain. (a) $C = 0$, (b) $C = 0.002$ (circles), and $C = 0.004$ (squares), (c) $C = 0.006$ (circles), $C = 0.008$ (diamonds), and $C = 0.014$ (squares).

strength is very weak, these three oscillators are phase synchronized, but the amplitudes are not correlated; moreover, there is a lag between the signals. For large coupling [Fig. 7.10(b)], there is still phase synchronization, but the lag has diminished and the amplitudes tend to be correlated. The transition from the scenario in Fig. 7.10(a) to the scenario in Fig. 7.10(b) is smooth and continuous: it resembles the transition from phase to lag synchronization studied in the above section.

Similar studies of phase synchronization in systems of N coupled non-identical chaotic Rössler oscillators, under different coupling schemes, and for different populations of oscillators can be found in the literature [Pikovsky et al. (1996); Osipov et al. (1997)]. In all these computer simulations, Rössler oscillators modified by the introduction of the new parameter ω, which controls the frequency of each oscillator were used. The populations of oscillators were then constructed by means of particular choices of

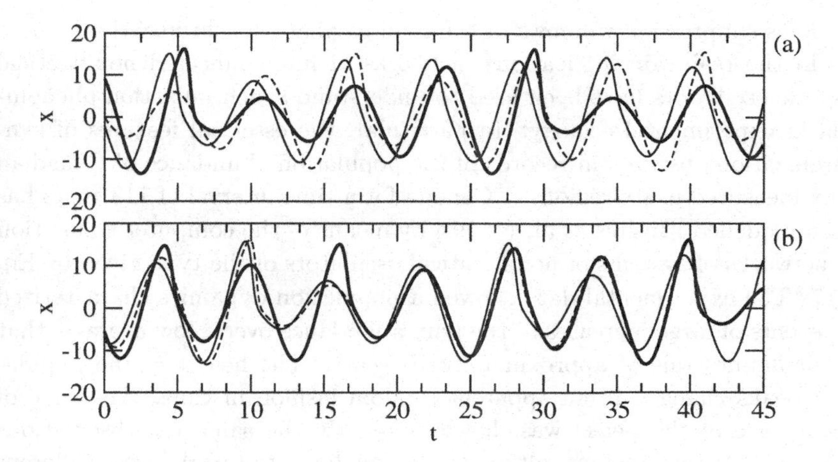

Fig. 7.10 Time series for the oscillators numbers 1 (thick line), 2 (dashed line), and 3 (thin line) of the chain at two coupling strengths (a) $C = 0.002$, and (b) $C = 0.014$.

the set of values of ω_i.

In a first series of numerical experiments [Pikovsky et al. (1996)], a large number of oscillators $N \sim 10^3$ where considered, with frequencies ω_i randomly chosen within a Gaussian distribution. The coupling scheme was of the mean field type, Eq. 6.20, acting on the differential equation for the x-variable. The value of the mean field $X = \sum_{j=1}^{N} \mathbf{x}_j/N$ was the measured observable of the system. The study of X as a function of the coupling strength revealed the continuous development of a macroscopic field as a result of a feedback process in which the synchronization of the phases of the oscillators increase the field, and this in turn enhances the synchronization of the phases.

When linear chains of $N \sim 10$ diffusively coupled oscillator with free boundary conditions were considered [Osipov et al. (1997)] the collapse of all the frequencies to a single common value was also observed. This was obtained for two kinds of structures of the population: (i) a linear distribution with the frequencies, ω_i, of each oscillator increased by a small amount, $\Delta\omega$, larger than that of its left nearest neighbor, and (ii) a random uniform distribution of frequencies in a narrow interval. In any case, the transition to phase synchronization occurred by means of clustering: groups of contiguous oscillators having the same frequency formed, which decreased in number and increased in size with the coupling strength until phase synchronization was achieved. These authors also reported the development of chaos suppression windows in the road to phase synchronization.

In the realm of applications, networks of interconnected nonidentical chaotic oscillators have been used to understand synchronization phenomena in experimental ecology. In particular, the essential features of synchronization observed in records of the population abundance of Canadian lynx measured in six regions of Canada for a time interval of 113 years has been modelled [Blasius et al. (1999)] by means of the computer simulation of networks of systems of non-identical oscillators of the type given by Eq. 7.17. The experimental data showed a population dynamics characterized by events of large increases in the lynx abundance over a low average, that occur at intervals of approximately 10 years. The height of the population peaks changed in an apparent random fashion in time. The time of occurrence of the peaks was, however, nearly the same for observations made in different regions, although the height of the peaks were different for each region. This phenomenology was interpreted as a phenomenon of phase synchronization in a spatially extended system. For this aim, the population in a single region was modelled by a three-dimensional chaotic oscillator whose variables are the local abundance of vegetation, herbivores and predators. These reproduced the features of the population dynamics just described for the individual observations in a single region. When two of these oscillators are mutually coupled, a transition from a desynchronized state to a lag synchronized state, through an intermediate state of phase synchronization, similar to that described in Subsection 7.1.3, was obtained. To study a spatially extended system, the dynamics of square lattices of $N \times N$ ($N = 20, 50, 100$) non-identical oscillator, with a random distribution of the parameters describing the rate of growth of herbivores was used. Each oscillator in the lattice was connected to first and second neighbors by diffusive terms describing the migration of herbivores and predators. The resulting dynamics showed phase synchronization of all the oscillators above a relatively low coupling strength in accordance to the spatial synchronization observed in the field.

Chapter 8

On the Dynamics of Coupled and Driven Chaotic Oscillators

This chapter is devoted to a summary of the main results presented in the book, to discuss their possible technical and scientific applications, and to speculate on the perspectives of the field both from the fundamental and the applied points of view.

8.1 An overview

This book is about the dynamics of coupled and driven chaotic oscillators, the main results presented are summarized in this section.

8.1.1 *Oscillators*

Oscillators are systems that have observables whose time evolutions remain bounded and change up and down around certain equilibrium value. Because this kind of behavior is often found in all the natural sciences, medicine and engineering, the study of the physics and mathematics of oscillators has become a fundamental scientific issue. A simple and meaningful type of oscillator is the harmonic oscillator which displays a periodic behavior. This is a linear system, whose theoretical study can be done exactly by standard mathematical techniques, and which sheds light on the basics of the nature of oscillations: the existence of stable equilibria around which the oscillations occur, the effect of dissipation as an initial condition erasing mechanism, and the necessity of forcing to have sustained oscillations.

In the most relevant applications of the physics of oscillators, these are not isolated entities, but subsystems in interaction with an environment. Two basic scenarios are relevant: oscillators being driven from their surroundings, and mutually coupled oscillators. The study of harmonic oscillators interacting with an environment results in two phenomena which have proven to be of great scientific and technical importance: resonance, when an oscillator is periodically driven, and normal modes when oscillators are mutually coupled.

Linearity, however, is only a particular condition fulfilled, even approximately, by a minority of systems in the natural world and in technology. Nonlinear oscillators are common, but its theoretical analysis is more difficult, and in many cases can be carried out only numerically. However, in view of the great impact that the study of linear oscillators has had, it appears that the investigation of the dynamics of nonlinear oscillators might be even more rewarding. Among the dynamic behaviors available to nonlinear oscillators there is one essentially different than the periodic, characteristic of linear oscillators, named chaos. This book has addressed the dynamics of coupled and driven chaotic oscillators; that is, the problem of the dynamics of chaotic oscillators in interaction with an environment.

The essential feature of chaos is sensitivity to initial conditions. This means that chaotic trajectories in phase space are unstable in the sense that any trajectory that starts in a point infinitely close to a chaotic trajectory will diverge exponentially from it. This implies that, because infinite precision in the measure of initial conditions is impossible, chaotic systems are unpredictable. Moreover, the time evolution of the dynamical variables that describe a chaotic system have to be aperiodic; because, otherwise its future state would be predictable. Chaotic trajectories in phase space are then complex entangled curves that never repeat themselves. The set of points visited by a trajectory of a chaotic system in its phase space belongs to an attractor which is a fractal sets of points whose characteristic feature is that it displays new details under successive magnification.

Invariant properties characterize chaotic attractors, and techniques to compute them have been developed. A fundamental one is the Lyapunov spectrum which measures and characterizes the overall instability of the chaotic orbits. The power spectra of chaotic signals are also significant properties that can be obtained for the sake of characterization. Unstable periodic orbits provide a fundamental approach to the analysis of chaotic attractors: a chaotic attractor has an embedded skeleton, made of an infinite set of unstable periodic orbits, which are continuously being visited for short intervals of time by the phase space trajectories. The less unstable periodic orbits of this skeleton can be obtained by appropriate procedures and relevant properties can be computed from them.

A given system, with dependence on environmental conditions or internal state, can behave chaotically or non chaotically. This means that the nature of the dynamics can change when control parameters of the system are changed. Bifurcation theory deals with these transitions, and several routes to chaos by which non-chaotic systems become chaotic have been identified; the period doubling route to chaos, and the route through quasiperiodicity are two of them.

For research purposes, chaotic systems are usually modelled by means of two basic types of mathematical entities: ordinary differential equations in a continuous time, named flows, or difference equations in a discrete time, named maps. Moreover, techniques have been developed to reconstruct the whole dynamics for experimental systems whose only known output is a measured scalar time signal. Be it by means of mathematical models, or by phase space reconstruction, the dynamics of chaotic systems can be studied, and characterized by the above mentioned properties as well as others.

8.1.2 *Driven chaotic oscillators*

A simple and meaningful form of interaction of a chaotic oscillator with its environment is when it is acted on by an external periodic force. Weakly driven chaotic oscillators have received considerable attention, and it has been observed that they may display two different interesting phenomena: the synchronization of the phase of the oscillator to the phase of the force, or the suppression of chaos for resonant frequencies. These share their interest as basic phenomena that may be observed in nature or in the laboratory, as well as used in potential technical applications.

The concept of the phase of a periodic oscillator can be generalized to a chaotic oscillator. There are several procedures that allow us to assign a phase to a chaotic oscillator, which describes its oscillatory dynamics around equilibrium points. These include the use of angles in an appropriate plane, analytic signals, stroboscopic plots, and return times. A relevant effect that may develop when a chaotic oscillator is acted on by a periodic force, whose oscillation has a well defined phase, is the locking of the phase of the oscillator to that of the force. This phenomenon has been observed in experiments made on chemical oscillators, plasma tubes, and electric circuits among other systems. It might have applications ranging from scientific, such as the study of biological rhythms or climatic phenomena, to technical, such as the technology of telecommunications.

The fact that a weak periodic force, having a resonant frequency, and an appropriate phase can suppress chaos, by turning an otherwise chaotic system periodic, is the other relevant effect of diving chaotic oscillators with periodic forces. Magneto-mechanical systems, discharge plasmas, and chemical reactions are among the systems where this phenomenon has been observed in the laboratory. Because periodic perturbations are common in the Earth and Life sciences, these appear as potential fields of application of this phenomenon. The possibility of suppression of chaos by means of weak perturbation is obviously of importance in many engineering fields were the irregular oscillations may be undesirable or harmful.

No less interesting is the interaction of a chaotic oscillator with its environment when the force implied evolves chaotically in time. In any case the force acting on the system of interest, has to be generated by another chaotic system. Then, in this case, there is an interaction setup in which a chaotic system works as a drive on another response chaotic system, by means of a chaotic signal going from the former to the later. Different mechanisms for driving chaotic systems with chaotic signals have been devised; these include continuous control, and variable substitution as the main types. The most relevant phenomenon studied until now is the synchronization of chaos, which is at least a curious result, given the sensitivity to initial conditions which is characteristic of chaotic systems.

Two basic situations are possible: when the drive and the response are identical, or nearly identical, and when they display definitely different individual dynamics. In the first case, synchronization is mainly understood as identical synchronization, which is characterized by the fact that the time evolution of the dynamical variables of the response is modified to exactly, or nearly exactly, match the time evolution of the correspondent dynamical variables of the drive. Moreover, special forms of synchronization can emerge for specific systems and coupling schemes; these include marginal and anticipated synchronization.

When the drive and the response are not identical systems, synchronization is still possible in the sense that the trajectory of the response is unambiguously determined by the trajectory of the drive. This is called generalized synchronization, and its development may not be obvious in many practical cases. Therefore, several techniques had to be invented for the purpose of detection. These include conditional Lyapunov exponents, auxiliary system approach, local neighborliness measures.

Electric circuits and different types of lasers have been the type of systems preferred by the experimentalists to successfully prove that all these forms of synchronization occur, beyond theory and computer simulations, in the real world. Applications in science and technology are promising and include the detection and understanding of synchronization phenomena in the brain, and the development of techniques to secure the privacy in communications, to name just a few.

An alternative to applying a perturbation, chaotic or periodic, to a chaotic oscillator and investigate what happens, is to design specific perturbations to be applied to a chaotic system with the aim of achieving a desired goal. This has been done to control chaos; i.e. to stabilize one of the unstable periodic orbits embedded in a chaotic attractor, chosen at will, by means of tiny perturbations. This is an important task because there are relevant cases when chaotic oscillations are undesirable, such as, for example, arrhythmias in medicine or harmful vibrations in engineering. To turn these chaotic oscillations into periodic is obviously interesting. Moreover, because there is an infinite number of different unstable periodic orbits in the attractor, such control techniques allow the use of chaotic oscillators as flexible multipurpose systems by means of the possibility of changing the particular periodic orbit that is stabilized. Two basic techniques have been devised to this aim: the OGY method, and the Pyragas method.

The OGY method is realized by means of small discrete perturbations applied to a system parameter according to a control law specifically defined for the system at hand. The method works on discrete maps, such as Poincaré maps, and has a solid theoretical background on the stability theory of the dynamics around unstable fixed points that have well defined stable and unstable directions. The idea applied to stabilize the dynamics is to use the control law to throw any phase space point that evolves in the neighborhood of the fixed point on top of its stable direction. The procedure is proven to work in a variety of systems which include magneto-mechanical devices, chemical reactions, electric circuits, lasers, fluids in motion, and biological systems.

The Pyragas method resorts to the application of a weak external continuous force to the chaotic oscillator. This method is also known as delayed feedback control. This is because the force is proportional to the difference between the actual value of a properly chosen observable of the system, and its value a time in the past equal to the period of the orbit to be stabilized. Once an unstable periodic orbit is stabilized the force becomes null. This technique has proven to be useful to stabilize orbits of low period in systems similar to those where the OGY method has worked.

8.1.3 *Mutually coupled chaotic oscillators*

Besides the case of a chaotic oscillator being driven from its environment there is a second case of interests. This is when a number of chaotic oscillators are in mutual interaction. This is interesting in many areas of science and technology where networks of interacting oscillating units are systems commonly found. Moreover, the study of continuous systems, where the understanding of spatial variations are relevant, leads to a problem of coupled chaotic oscillators when an approach based on discretization is made.

Even the simplest case, when two identical oscillators are mutually coupled is interesting. This is because this is a configuration often found in practice, and because this represents a basic case which displays many of the phenomenology that is observed in more complex networks. Asymptotically stable identical synchronization between the two oscillators happens to be the basic phenomenon that is observed. Moreover, a variety of other dynamical behaviors, such as periodicity, quasi-periodicity, chaos, and multistability may develop as a result of the coupling. The synchronized state and these complex phenomena have been observed in experiments made with electric circuits and lasers.

When more than two identical oscillators are coupled, the same phenomenology described in the above paragraph for two oscillators can be observed. Moreover, this phenomenology is enriched with the possibility of development of partial synchronization of chaos. In this case the whole system of many oscillators becomes divided in subsets, in which the oscillators evolve in identical synchrony between them, but desynchronized from the oscillators of the other subsets. Complete synchronization is used to design the synchronized state in which all the oscillators evolve in identical synchrony, in contrast with what happens in the partially synchronized state. This last phenomenon is a mechanism by which spatial patterns can emerge in a system composed of several, or many, interacting identical chaotic oscillators. Partial synchronization of chaos has been observed in the laboratory, in particular in investigations made with electric circuits, and in chemical reacting systems.

When the oscillators that are coupled are different enough so as to display different individual dynamics, identical synchronization is only possible in the limit of strong coupling. For weak coupling several forms of synchronization weaker than identical synchronization may develop. Again the system composed of two oscillators, despite its simplicity, is enough to study some of the basic forms of synchronization available to non-identical mutually coupled chaotic oscillators.

For very mild coupling some systems display amplitude envelope synchronization. This is a faint form of synchronization in which, neither the amplitudes, nor the phases of the dynamic variables that describe the oscillators, are correlated. This form of synchronization is between the envelopes of these variables. That is, the maxima and minima that these variables can attain at each time are correlated between the two systems. The mechanism for amplitude envelope synchronization, as shown by the spectral analysis, is a feedback process between the two oscillators in which motions at small frequencies are mutually excited. This form of synchronization, for now, has been observed only in computer experiments mainly made on models of electric circuits; a potential application is the study of synchronization of extreme phenomena; i.e., those that correspond to peaks of certain dynamic variables.

Phase synchronization develops in mutually coupled chaotic oscillators under mild coupling. It is essentially the same phenomenon of phase synchronization that appears when a chaotic oscillator is driven by a periodic force. In this case, the phases of the two oscillators become modified to evolve in pace with each other. The amplitudes of the variables that describe the dynamics of the oscillators, however remain uncorrelated. The transition from the desynchronized state to the phase synchronized state is not a threshold one, but a gradual development of a locked state. This is appreciated when spectral analysis and Lyapunov exponents are used to monitor the transition, which develops along an intermittence route. Chemical oscillators and nonlinear optical interferometers are among the systems where phase synchronization has been observed in the laboratory.

Beyond phase synchronization, for further increase of the coupling strength, a stronger form of synchronization, known as lag synchronization, may develop. In this case the variables that describe the time evolution of the two oscillators evolve in identical synchronization, but for a time lag between them. This time lag decreases when the coupling strength increases; then, for increasing coupling strength, lag synchronization converges to identical synchronization. When there are large differences between the two oscillators lag synchronization becomes generalized lag synchronization, where the word generalized has to be understood in the sense of generalized synchronization discussed above. Time lagged synchronization, in the identical and in the generalized forms, has been observed in experiments made on electric circuits.

The dynamics of systems of more than two mutually coupled oscillators combines all the above elements. For weak coupling, a system of N different oscillators may reach a state of complete phase synchronization with all the phases evolving in pace, while the amplitudes are uncorrelated. The transition, however, is through a process of partial phase synchronization of chaos, in which clusters made of subsystems of oscillators in synchrony within each cluster, but without synchrony between different clusters.

The account of synchronization and control of chaos presented above is just an introduction to the basic results of scientific research made by many scientists mainly in the last fifteen years. By now, the amount of knowledge accumulated, and the level of maturity reached have gained recognition for the synchronization and control of chaos as a specific field of research by the scientific community. This is still an alive, and promising one, as demonstrated by the activity developed in the last few years which, which besides an important amount of research papers appearing in regular scientific journals, has produced special publications in the form of proceedings of international meetings [Pérez-Muñuzuri et al. (1999); Boccaletti et al. (2001); Kurths et al. (2003)], special issues on control of chaos [Kapitaniak (1997); Arecchi et al. (1998)], special issues on chaos synchronization [Kurths (2000)], and special issues on control and synchronization of chaos [Kennedy and Ogorzalek (1997); Ditto and Showalter (1997); Chen and Ogorzalek (2000); Kapitaniak (2003)] that have appeared in several specialized journals.

This suggests that new fundamental knowledge in this field is still to come in many possible forms. This may include the discovery of new phenomena resulting from the coupling and driving of chaotic oscillators, which will contribute to expand the field. The study of more complex systems than the simple low dimensional chaotic flows, that have received preferential attention in the past, appears promising too. These systems might be networks made of many oscillators, high dimensional systems, and even special cases of low order systems bearing complicated dynamics. From a fundamental point of view, it appears that some kind of theoretical unifying scheme for the variety of synchronization and chaos suppression phenomena, that have been discovered, is becoming a necessity. These types of unification, that have been made in other scientific fields in the past, happen to be useful for a better understanding of the field, and for the new lines of research that they suggest. Finally, the application of the knowledge achieved to the solution of problems in science and technology appears as a must that will result in scientific and practical benefits.

8.2 On the scientific and technical applications of synchronization and control

Most of the research on the dynamics of driven and coupled chaotic oscillators, that has led to the rise of the field of synchronization and control of chaos, has been made by mathematicians and physicists. Concepts and tools have been developed and experimentally tested. Most of these tests have been made on mechanical devices, electric circuits and lasers; also, chemical oscillators have often been used. The amount of knowledge accumulated suggests that the time for applications has come. The use of this knowledge to better understand systems and phenomena of interest in the natural sciences, and to work out solutions to problems in medicine and engineering would give a new sense to all that has been achieved. Because of this, this book will end with the discussion of work done on some major problems in the Life sciences, the Earth sciences, and engineering, where material studied in this book, or developed from it, has been used to contribute to deal with these problems. These are aimed to be examples that illustrate the contributions that synchronization and control of chaos could make in all the sciences, medicine and technology.

8.2.1 *Synchronization in neurobiology*

The nervous system is a complex, differentiated, and organized network made of a large number of neurons interconnected by means of synapses. A fundamental problem is implied in the task of understanding how the nervous system works, this is known as the problem of the large scale integration in the brain [Varela et al. (2001)]. The fundamental constitutive element of the nervous system, and in particular of the brain, is a single neuron, whose electrical activity provides the basic mechanism for neurobiological activity. A neuron receives chemical inputs from other neurons through its dendrites, integrates these inputs to produce electrical impulses that propagate along the axon, which finally result in an output to other neurons through the synapses. The electrical activity is measurable, and has an oscillatory nature like the bursting-spiking dynamics described in Chapter 2 for the Hindmarsh–Rose neuron model [Hindmarsh and Rose (1984)]. Moreover, neurons are specialized in the sense that, depending on their position in the nervous systems, they bear special properties and configure specific differentiated structures by means of connections to other neurons. In particular, the brain is organized in functionally specialized areas, which have its own location, and perform specific tasks (cognitive, sensitive, motor). These specialized areas have differentiated parts or structures within. The realization of any cerebral activity implies the coordinate performance of several tasks, that occur in different and possibly distant specialized areas of the brain, and the rest of the nervous system. Therefore the correct working of the nervous system implies the integration of the neural activity in several scales: (i) the microscale where the interaction between two, or few, neurons occurs, (ii) the mesoscale of a specialized area which has a width of the order of one centimeter or less, and (ii) the macroscale between different areas (with a scale of the order of several centimeters). Under these circumstances it is pertinent to ask how all these structures can work together to produce a single elementary activity such as the integral perception of an object or a motor action: this is the problem of the large scale integration in the brain [Varela et al. (2001)].

Several authors have proposed synchronization as the basic physical mechanism that allows this large scale integration [Damasio (1990); Mesulam (1990); Varela et al. (2001)]: each elementary cerebral activity is made up by collectives of neurons whose electrical activity oscillates synchronously to produce this event in the different scales of the nervous system. The detection and characterization of this synchrony is then of major interest because, if such synchrony exists, an approach to deal with the problem of the large scale integration would be opened. This will allow us to use neurons, or specialized areas, modelized as nonlinear oscillators. Then, all the concepts and techniques on chaos synchronization presented in this book will be potentially useful tools to deal with this problem at any of the scales said above. It must be noted that, in many cases, this is not an easy problem because the experimental observations used are noisy and nonstationary. However, the concepts and techniques introduced to study the different forms of synchronization provide the basics to guess the nature of the synchronization processes that are happening at the different neurobiological levels, and then to devise additional appropriate techniques to perform effective detections of synchronization events. Some examples of work of this kind will be discussed now.

In the microscopic scale, the dynamics of two lobster neurons mutually coupled by their natural synapses and by artificial electric synapses, and isolated from the rest of the tissue by chemical blocking and photoinactivation of the synapses to other neurons has been the object of experimental study [Elson et al. (1998)]. A single isolated neuron was able to exhibit periodic and chaotic spiking and bursting dynamics qualitatively similar to those sketched in Fig. 2.10. The synchronization phenomena were mainly studied in the bursting regime where the dynamics occurs in two time scales, the low frequency scale of the bursts and the high frequency scale of the spikes. Synchronization as a function of the intensity of the coupling between the neurons was then studied separately for each time scale. Using a filter to remove the high frequency part of the signals it was possible to observe phenomena of identical, or nearly identical, synchronization in the bursting rhythms by means of the use of parametric plots. Moreover, an analysis of the times of occurrence of the spikes was used to identify phase differences and lags between the signals, this allowing to observe phase synchronized behavior with a certain lag. Three synchronization regimes were identified: desynchronization, synchronization of the burst without synchronization of the spikes, synchronization of burst and spikes.

Synchronization in the macroscale in the human brain has been studied using techniques constructed on the ground of the concept of generalized synchronization [Breakspear and Terry (2002)]. The aim of these experiments was to detect nonlinear interdependence between different regions of the brain obtained by scalp electroencephalography (EEG) of healthy individuals. Pairs of EEG signals obtained from different regions of the brain were considered as coming from two different oscillators, whose phase space dynamics was reconstructed by standard embedding techniques; then, generalized synchronization was considered as an appropriate conceptual base to develop such analysis. The quantity used to detect synchronization in this case was a parameter known as normalized future prediction error [Terry and Breakspear (2003)], whose definition is based on the property of generalized synchronization which allows prediction of the state of one, of two coupled oscillator, from the state of the other. Using this technique, these authors [Breakspear and Terry (2002)] were able to report detection of nonlinear interdependence between different brain regions.

Phase synchronization in neurobiological systems has been studied in the macroscopic scale [Tass el al. (1998)] by looking for phase synchronization between signals from two different areas of the brain obtained by magnetoencephalographic techniques. The properties of the signals so obtained happened to be irregular, nonstationary and noisy. Therefore the techniques of detection of phase synchronization presented in Chapters 3 and 7, were unable to detect epochs of synchronization, and had to be adapted to deal with this more complex situation. The detection of phase synchronization in these cases was based on the observation of statistical regularities when two signals are compared [Tass el al. (1998)]. By the application of these statistical techniques to the study of a muscular tremor event in a patient of Parkinson disease these authors were able to detect 1 : 1 phase synchronization between two different brain areas implied in the process. Moreover, comparing signals from electroencephalography with signals of electromyography, from the muscles implied in the tremor, it was also possible to detect 1 : 2 neuromuscular phase synchronization.

8.2.2 *Synchronization in the Earth sciences*

Deterministic chaos is being recognized as a useful tool that can be useful to describe and understand the complexity inherent to the dynamics of the atmosphere and the climate [Tsonis (2001)]. Such recognition is linked to the identification of patterns of atmospheric circulation that have a large spatial scale, a high degree of persistence in time, and are described by observables whose time evolutions are aperiodic [Barnston and Livezey (1987); Tsonis (2001)]. One example of such structures is the North Atlantic Oscillation (NAO), which consists of the alternation of two spatial patterns of pressure along the North Atlantic, one with higher pressures in the arctic latitudes and the other with lower pressures in that area. Another example is the el Niño/South oscillation (ENSO), in which an atmospheric pattern develops over the South Pacific Ocean with high pressure on the West and low pressures on the East, linked to an anomalous warming of the sea near the costs of South America. Moreover, there is experimental evidence that proves that atmospheric-ocean structures of this kind can modify the climatic condition in nearby, and even far, places in the globe, which indicates that they cannot be considered as isolated systems.

The identification of such subsystems within the climate system, and their connectivity suggest a picture in which networks of nonlinear oscillators in mutual interaction would provide idealized, but useful models for many climate and weather events [Duane et al. (1999); Tsonis (2001)]. In such models the different synchronization scenarios studied in this book would play an important role.

It is expected that low dimensional models for atmospheric and oceanic structures may be useful tools for such modelization. They serve to fill the gap between qualitative reasoning and very complex, and resource consuming, modelization. Their advantages are their relative simplicity, which allows wide scope explorations of the oscillator parameter space, still being quantitative and allowing complex nonlinear dynamics. One example is the Lorenz model for convective motion in fluids [Lorenz (1963)] which has been discussed in Chapter 2. Another example is the Lorenz general atmospheric circulation model [Lorenz (1984)], which describes the circulation of the atmosphere around the globe by means of three variables, one that measures the intensity of the westerly air current circling the globe, and two more that describe the dynamics of the large scale eddies superposed to that current. These elements interact with each other, and the model allows to obtain its combined dynamics under external force, and internal dissipation, which are included as system parameters. This model has been used, for example, in studies of predictability of the atmospheric circulation dynamics [Shil'nikov et al. (1995); González-Miranda (1997)].

One possible use of the ideas on synchronization is the detection of synchronized motion to show up the synchronized behavior between different points in a pattern of atmospheric circulation. This is useful to define and characterize this pattern. Because different points in the pattern have different properties, the oscillations of climatic or meteorological variables have to be different in different points; therefore, forms of synchronization such as generalized synchronization and phase synchronization have to be considered. In an investigation of this type, pairs of time series of daily temperature records pertaining to different European locations (Oxford and Vienna), and temperature and precipitation records at the same place (Oxford), have been tested for phase lagged synchronization [Rybski et al. (2003)]. The analytic signal approach was used to determine the phase of each scalar time series; moreover, because of the noisy nature of the experimental data, specific techniques devised to detect phase synchronization from noisy data sets [Tass el al. (1998)] had to be used. The results obtained showed unambiguously the existence of phase synchronization, with a certain lag, for distant pairs of locations for which, however, a meteorological connection was expected. Synchronization between the temperature and the precipitation at the same place was also observed in this way. Moreover, the tests based on the search for lagged phase synchronization were compared with traditional techniques, based on the calculation of cross correlations, finding that the former provided sharper indications of the existing connections between the time series considered.

Modelling of climatic and meteorological scenarios is another application of interest. Coupled and driven low dimensional chaotic flows have been used [Brindley et al. (1995); Stefanski et al. (1996)] to study the interaction between an extratropical circulation structure with high turbulent behavior, i.e. an intense storm, with other structures of extratropical circulation having a less turbulent character, i.e. anticyclonic regions. In particular, Lorenz oscillators [Lorenz (1963)] tuned to different parameter values were used to model the two different structures, and a diffusive coupling was considered. Synchronized behaviors were observed and interpreted as phenomena in which atmospheric structures in a region, can modify the dynamics of structures in other regions, even by means of a weak interaction. For milder couplings, for which synchronization was not achieved, coupling resulted in a decrease of the chaos of the highly turbulent structure, and then interpreted by the authors as a predictability enhancing mechanism that might explain the existence of situations of increased predictability when more turbulent conditions were expected.

The dynamics of two coupled low order oscillators describing, one of them, the general circulation of the atmosphere in the Northern Hemisphere [Lorenz (1984)], and the other the circulation of water in the North Atlantic Ocean driven by temperature and salinity [Roebber (1995)], known as thermohaline circulation, has been carried out to investigate the effect of the coupling between atmosphere and ocean in the climate variability. In this case, the two coupled oscillator were, not only parametrically, but even structurally different. Moreover, their characteristic response times to perturbations were very different. The dynamics behavior of the coupled system, then, happened to be quite complex: changes in the circulation of the atmosphere cause the change in the state of the ocean, which in turn results in a change in the circulation behavior of the atmosphere; i.e. the dynamics that result is a succession of adjustments, or attempts of synchronization, of one oscillator to another that occur in different time scales. The overall scenario is then one in which different phases of the ocean oscillations have associated different patterns of the general circulation of the atmosphere. This behavior mimics transition between different climates observed during the last 100,000 years [Roebber (1995)].

8.2.3 *Chaotic communications*

The irregular nature of the dynamics of chaotic systems, combined with the possibilities of synchronize or control chaos have triggered a line of research in the area of applied chaos aimed to use these properties of chaos to enhance privacy in telecommunications. Privacy means security; i.e. transmission of messages, between a transmitter and a receiver, in such a way that this message is inaccessible to possible eavesdroppers.

A straightforward implementation of a chaotic secure communication system is based on identical synchronization of chaos, and is known as signal masking and recovery [Cuomo and Oppenheim (1993a)]. This means to send a low amplitude message signal, $s(t)$, hidden within a large amplitude chaotic carrier, $X(t)$. The transmitter and the receiver have to contain two identical chaotic systems, coupled unidirectionally from the transmitter to the receiver, and such that the response, in the receiver, is able to synchronize to the drive, in the transmitter, in the sense of identical synchronization, when driven by $X(t)$. That means that the conditional Lyapunov exponents of the response have to be negative. In this case, the structural stability of identical synchronization, which implies robustness of the synchronized state against noise added to the signal, and against system parameter mismatch is essential for the recovery of the message. In the transmitter, the output signal of the chaotic system, $X(t)$, is modified by adding the signal which contains the message, so that what is sent is $X(t) + s(t)$, with $s(t)$ so weak, compared to the $X(t)$, as to be equivalent to a noise affecting to it. Then the message, $s(t)$, has been masked by the carrier $X(t)$. At the receiver the input signal is separated in two copies, one is used to drive the response and create a copy of the chaotic carrier, $X'(t)$, which has to be in nearly identical synchrony with $X(t)$; i.e. $X'(t) \approx X(t)$. Then $X'(t)$ is subtracted to the other copy, $X(t) + s(t)$, to recover the the message: $[X(t) + s(t)] - X'(t) \approx s(t)$.

A successful experimental realization of signal masking and recovery was first made using electric circuits [Cuomo and Oppenheim (1993a)]. The chaotic system at the core of the transmitter and the receiver was an electric analog circuit implementation of the Lorenz model working in the chaotic regime. The x signal of this system was the chaotic carrier used to mask a continuous speech signal that was later fairly recovered at the receiver, then proving the feasibility of this technique.

Experimental realization of this masking scheme in an optical system has also been reported [VanWiggeren and Roy (1998)]. Two identical Erbium-doped fiber ring lasers were used as chaotic oscillators in the transmitter and the receiver, and an optical fiber line was used to connect them. A technical advantage of this system is the possibility of working with fast chaotic signals working at large bandwidths (of several hundreds of MHertz). In the experiment, a square wave signal working in the range of ten megahertz was added to the broadband chaotic electric field of the drive laser and transmitted. The message was recovered in the receiver after subtracting the synchronized field of the response from the input signal, and low-pass filtering the difference signal. As a bonus, it was observed that the nonlinearities in the lasers worked in such a way that the frequency components of the message were preferentially amplified in the receiver [VanWiggeren and Roy (1998)]. This allowed the use of smaller amplitudes of the message signal in the transmitter, and then to improve the degree of security.

Control of chaos techniques have also been used for the transmission of messages by means of chaotic signals [Hayes et al. (1993); Hayes et al. (1994)]. In this approach the message to be sent is previously coded in binary form as a sequence of 0s and 1s. The transmitter is a chaotic oscillator whose dynamics has been studied previously to choose a criteria to assign 0s and 1s to some feature of the dynamics. For example, for attractors where the motion contains jumps between two lobes, like the Lorenz model or the circuit of Chua, the binary coding is made by assigning a 0 to one lobe and 1 to the other [Hayes et al. (1993)]; although, other criteria can be used with attractors having other topologies [Hayes et al. (1994)]. This allows us to assign a symbolic sequence to a chaotic trajectory which is defined by the series of 0s and 1s that result according to that criteria. To implement the communication protocol, a first phase has to be developed where the relation between the crossing points on a surface of section properly chosen and the symbolic sequences that follow to each crossing has to be established. This is then used to define a control law for the sequences of 0s and 1s to be followed by the system dynamics. Once this control law has been defined, a second phase is entered, and the binary message is sent as a chaotic signal whose symbolic sequence is precisely the sequence of 0s and 1s that codify the message. Assuming that the receiver knows the criteria used to define the symbolic dynamics, no synchronizing chaotic system is needed, just the observation of the symbolic sequence of the signal is enough to recover the message.

With these examples, this book, about coupled and driven chaotic oscillators, comes to an end. The research discussed in this section, picked up from a variety of sources, is aimed to exemplify what might be a major line of development of this field in the next few years: the application of the new ideas and techniques in all the sciences, medicine and engineering. These will probably evolve together with further development within the field, as discussed in the above section. If this book is, somehow, helpful to those willing to engage in either, or both, of these applied or fundamental enterprises the time and energy spent in writing it will have been worthwhile.

Bibliography

Abarbanel, H. D. I., Brown, R., Sidorowich, J. J. and Tsimring, L. Sh. (1993). The analysis of observed chaotic data in physical systems, *Rev. Mod. Phys.* **65**, pp. 1331-1392.

Abarbanel, H. D. I., Rulkov, N. F. and Sushchik, M. M. (1996). Generalized synchronization of chaos: the auxiliary system approach, *Phys. Rev. E* **53**, pp. 4528-4535.

Afraimovich, V. S., Verichev, N. N. and Rabinovich, M. I. (1986). Stochastic synchronization of oscillations in dissipative systems, *Izv. VUZ. Radiofiz. RPQAEC* **29**, pp. 795-803.

Arecchi, F. T., Boccaletti, S., Ciofini, M., Meucci, R. and Grebogi, G. (1998). The control of chaos: theoretical schemes and experimental realizations, *Int. J. Bifurcation and Chaos* **8**, pp. 1643-1655.

Argoul, F., Arneodo, A. and Richetti, P. (1987) Experimental evidence for homoclinic chaos in the Belousov-Zabotinskii reaction. *Phys. Lett. A* **120**, pp. 269-275.

Arneodo, A., Coullet, P. and Tresser, C. (1981). Possible new strange attractors with spiral structure, *Commun. Math. Phys.* **79**, pp. 573-579.

Atlee Jackson, E. *Perspectives of Nonlinear Dynamics* (Cambridge University Press, New York, 1991), Vol. 1, pp. 41-44.

Auerbach, D., Cvitanovic, P., Eckmann J.-P., Gunaratne, G. and Proccacia, I. (1987). Exploring chaotic motion through periodic orbits, *Phys. Rev. Lett.* **58**, pp. 2387-2389.

Badii, R., Brun, E., Finardi. M., Flepp, L., Holzner, R., Parisi, J., Reyl, C. and Simonet, J. (1994). Progress in the analysis of experimental chaos through periodic orbits, *Rev. Mod. Phys.* **66,** pp. 1389-1415.

Barnston, A. G. and Livezey, R. E. (1987). Classification, seasonality and persistence of low-frequency atmospheric circulation patterns. *Mon. Weather Rev.* **115**, pp. 1083-1126.

Bassett, M. R. and Hudson, J. L. (1988) Shil'nikov chaos during copper electodissolution. *J. Phys. Chem.* 92, pp. 6963-6966.

Benettin, G., Galgani, L., Giorgilli, A. and Strelcyn, J.-M. (1980a). Lyapunov characteristic exponents for smooth dynamical systems and for hamiltonian systems; a method for computing all of them. Part 1: Theory, *Meccanica* **15**, pp. 9–20.

Benettin, G., Galgani, L., Giorgilli, A. and Strelcyn, J.-M. (1980b). Lyapunov characteristic exponents for smooth dynamical systems and for hamiltonian systems; a method for computing all of them. Part 2: Numerical application., *Meccanica* **15**, pp. 21–30.

Bindu, V. and Nandakumaran, V. M. (2000). Numerical studies on bidirectionally coupled directly modulated semiconductor lasers, *Phys. Lett. A* **277**, pp. 345–351.

Boccaletti, S., Valladares, D. L., Kurths, J., Maza, D. and Mancini, H. (2000). Synchronization of chaotic structurally nonequivalent systems, *Phys. Rev. E* **61**, pp. 3712-3715.

Boccaletti, S., Burguete, J., González-Viñas, W., Mancini, H. L. and Valladares, D. L. (2001). Editorial, *Int. J. Bifurcation and Chaos* **11**, pp. 2529-2350.

Bracikowski, C. and Roy, R. (1991). Chaos in a multimode solid-state laser system, *Chaos* **1**, pp. 49-64.

Braiman, Y. and Goldhirsch, I. (1991). Taming chaotic dynamics with weak periodic perturbations, *Phys. Rev. Lett.* **66**, pp. 2545–2548.

Breakspear, M. and Terry, J. R. (2002). Detection and description of non-linear interdependence in normal multichannel human EEG data, *Clin. Neurophysiol.* **113**, pp. 735–753.

Brindley, J., Kapitaniak, T. and Kocarev, L. (1995). Controlling chaos by chaos in geophysical systems, *Geophys. Res. Lett.* **22**, pp. 1257-1260.

Blasius, B., Huppert, A. and Stone, L. (1999). Complex dynamics and phase synchrony in spatially extended ecological systems, *Nature* **399**, pp. 354-359.

Candaten, M. and Rinaldi, S. (2000). Peak to peak dynamics: a crytical survey, *Int. J. Bifurcation and Chaos* **10**, pp. 1805–1819.

Cao, L.-Y. and Lai, Y.-C. (1998). Antiphase synchronism in chaotic systems, *Phys. Rev. E.* **58**, pp. 382-386.

Carroll, T. L. and Pecora, L. M. (1991). Synchronizing chaotic circuits. *IEEE Trans. Circuits and Systems*, **CAS-38**, pp. 453–456.

Casdagli, M., Eubank, S., Farmer, J. D. and Gibson, J. (1991). State space reconstruction in the presence of noise, *Physica D* **51**, pp. 52-98.

Chacón, R. (1995). Suppression of chaos by selective resonant parametric perturbations, *Phys. Rev. E.* **51**, pp. 761-764.

Chacón, R. (1996), Geometrical resonance as a chaos eliminating mechanism, *Phys. Rev. Lett.* **77**, pp. 482-485.

Chacón, R., Palmero, F. and Balibrea, F. (2001). Taming chaos in a driven Josephson junction, *Int. J. Bifurcation and Chaos* **11**, pp. 1897–1909.

Chen, G. and Ogorzalek, M. J. (2000). Editorial, *Int. J. Bifurcation and Chaos* **10**, pp. 509–509.

Chen, J. Y., Wong, K. W. and Shuai, J. W. (2001). Properties of phase locking with weak phase-coherent attractors, *Phys. Lett. A* **285**, pp. 312-318.

Chua, L. O., Kocarev, L., Eckert, K. and Itoh, M. (1992). Experimental chaos synchronization in Chua's circuit, *Int. J. Bifurcation and Chaos* **2**, pp. 705–708.

Chua, L.O., Wu, C. W., Huang, A. and Zhong, G.-Q. (1993). An universal circuit for studying and generating chaos-Part II. Strange Attractors, *IEEE Trans. Circuits and Systems*, **CAS-40**, pp. 745–761.

Colet, P. and Braiman, Y. (1996). Control of chaos in multimode solid state lasers by the use of small periodic perturbations, *Phys. Rev. E.* **53**, pp. 200-206.

Crawford, J. D. (1991). Introduction to bifurcation theory, *Rev. Mod. Phys.* **63,** pp. 991-1037.

Cuomo, K. M. and Oppenheim, A. V. (1993a). Circuit implementation of synchronized chaos with applications to communications, *Phys. Rev. Lett.* **71**, pp. 65-68.

Cuomo, K. M. and Oppenheim, A. V. (1993b). Synchronization of Lorenz-based chaotic circuits with appplications to communications, *IEEE Trans. Circuits and Systems*, CAS-40, pp. 626–633.

Cvitanovic, P. (1988). Invariant measurement of strange sets in terms of cycles. *Phys. Rev. Lett.* **61**, pp. 2729-2732.

Damasio, A. (1990). Synchronous activation in multiple cortical areas: a mechanism for recall. *Sem. Neurosci.* **2**, pp. 287–296.

de Sousa Vieira, M. and Lichtenberg, A. J. (1996). Controlling chaos using feedback with delay. *Phys. Rev. E* **54**, pp. 1200-1207.

del Rio, E., Velarde, M. G., Rodríguez-Lozano, A., Rul'kov, N. F. and Volkovskii, A. R. (1994). Experimental evidence for synchronous behavior of chaotic nonlinear oscillators with unidirectional or mutual driving, *Int. J. Bifurcation and Chaos* **4**, pp. 1003-1009.

Ding, W. X., She, H. Q., Huang, W. and Yu, C. X. (1994). Contolling chaos in a discharge plasma, *Phys. Rev. Lett.* **72**, pp. 96–99.

Ding, M. and Ott, E. (1994). Enhancing synchronism of chaotic systems, *Phys. Rev.* E **49,** pp. R945-R948.

Ditto, W. L., Rauseo, S. N. and Spano, M. L. (1990). Experimental control of chaos, *Phys. Rev. Lett.* **65**, pp. 3211–3214.

Ditto, W. L. and Showalter, K. (1997). Introduction: control and synchronization of chaos, *Chaos* **7**, pp. 509-511.

Duane, G. S., Webster, P. J. and Weiss, J. B. (1999). Co-occurrence of northern and southern hemisphere blocks as partially synchronized chaos, *J. Atmos. Sci.* **56**, pp. 4183-4205.

Eckmann, J.-P. (1981). Roads to turbulence in dissipative dynamical systems, *Rev. Mod. Phys.* **53,** pp. 643-654.

Eckmann, J.-P. and Ruelle, D. (1985). Ergodic theory of chaos and strange attractors, *Rev. Mod. Phys.* **57,** pp. 617-656.

Elaydi, S. N., *Discrete chaos* (Chapman and Hall/CRC, Boca Raton, FL, 2000).

Elson, R. C., Selverston, A. I., Huerta, R., Rulkov, N. F., Rabinovich, M. I. and Abarbanel, H. D. I. (1998). Synchronous behavior of two coupled biological neurons, *Phys. Rev. Lett.* **81**, pp. 5692-5695.

Farmer, J. D. (1982). Chaotic attractors of an infinite-dimensional dynamical system, *Physica D* **4**, pp. 366-393.

Feigenbaum, M. J. (1978). Quantitative universality for a class of nonlinear transformations, *J. Stat. Phys.* **19**, pp. 25-52.

Feigenbaum, M. J. (1983). Universal behavior in nonlinear systems, *Physica D* **7**, pp. 16-39.

Franceschini, V., Giberti, C. and Zheng, Z. (1993). Characterization of the Lorenz attractor by unstable periodic orbits, *Nonlinearity* **6**, pp. 251-258.

Fronzoni, L., Giocondo, M. and Pettini, M. (1991). Experimental evidence of suppression of chaos by resonant parametric perturbations, *Phys. Rev. A* **43**, pp. 6483–6487.

Fujisaka, H. and Yamada, T. (1983). Stability theory of synchronized motion in coupled-oscillator systems, *Prog. Theor. Phys.* **69**, pp. 32–47.

Gabor, D. (1946). Theory of Communication, *J. IEE (London)* **93(III)**, pp. 429-457.

Garfinkel, A., Spano, M. L., Ditto, W. L. and Weiss, J. N. (1992). Contolling cardiac chaos, *Science* **257**, pp. 1230-1235.

Gaspard. P. and Nicolis, G. (1983). What can we learn from homoclinic orbits in chaotic dynamics?. *J. Stat. Phys.* **31**, pp. 499–518.

Goldstein, H., Poole, C. and Safko, J. *Classical Mechanics, 3rd ed.* (Addison Wesley, San Francisco, CA, 2002).

Gomes, M. G. M. and King, G. P. (1992). Biestable chaos. II. Bifurcation analysis, *Phys. Rev.* A **46,** pp. 3100-3110.

González-Miranda, J. M. (1996a). Chaotic systems with a null conditional Lyapunov exponent under nonlinear driving, *Phys. Rev. E* **53**, pp. R5–R8.

González-Miranda, J. M. (1996b). Synchronization of symmetric chaotic systems, *Phys. Rev. E* **53**, pp. 5656-5669.

González-Miranda, J. M. (1997). Predictability in the Lorenz low-order general atmospheric circulation model, *Phys. Lett. A* **233**, pp. 347-354.

González-Miranda, J. M. (1998a). Amplification and displacement of chaotic attractors by means of unidirectional chaotic driving, *Phys. Rev. E* **57**, pp. 7321–7324.

González-Miranda, J. M. (1998b). Using continuous control for amplification and displacement of chaotic signals, *Eur. Phys. J. B* **6**, pp. 411-418.

González-Miranda, J. M. (1999). Amplification phenomena under chaotic driving in a model of chemical chaos, *Int. J. Bifurcation and Chaos* **9**, pp. 2237–2242.

González-Miranda, J. M. (2002a). Amplitude envelope synchronization in coupled chaotic oscillators, *Phys. Rev. E* **65**, pp. 036232-(1-9).

González-Miranda, J. M. (2002b). Phase synchronization and chaos suppression in a set of two coupled nonlinear oscillators, *Int. J. Bifurcation and Chaos* **12**, pp. 2105–2122.

González-Miranda, J. M. (2002c). Generalized synchronization in directionally coupled systems with identical individual dynamics, *Phys. Rev. E* **65**, pp. 047202-(1-4).

González-Miranda, J. M. (2003). Observation of a continuous interior crisis in the Hindmarsh–Rose neuron model, *Chaos* **13**, pp. 845-852.

Grebogi, C., Ott, E. and Yorke, J. A. (1982). Chaotic attractors in crisis, *Phys. Rev. Lett.* **48**, pp. 1507-1510.

Grebogi, C., McDonald, S. W., Ott, E. and Yorke, J. A. (1983). Final state sensitivity: an obstruction to predictability, *Phys. Lett. A* **99**, pp. 415-418.

Grebogi, C., Ott, E. and Yorke, J. A. (1988). Unstable periodic orbits and the dimensions of multifractal chaotic attractors. *Phys. Rev. A* **37**, pp. 1711-1724.

Guckenheimer, J. and Holmes, P., *Nonlinear Oscillations, Dynamical Systems, and Bifurcation of Vector Fields* (Springer Verlag, New York, 1983).

Guderian, A., Münster, A. F., Kraus, M., and Schneider, F. W. (1998). Electrochemical chaos control in a chemical reaction: experiment and simulation, *J. Phys. Chem. A* **102**, pp. 5059–5064.

Güémez, J. and Matías, M. A. (1995). Modified method for synchronizing and cascading chaotic systems, *Phys. Rev.* E **52,** pp. R2145-R2148.

Hayes, S., Grebogi, C. and Ott, E. (1993). Communicating with chaos, *Phys. Rev. Lett.* **70**, pp. 3031-3034.

Hayes, S., Grebogi, C., Ott, E. and Mark, A. (1994). Experimental control of chaos for communication, *Phys. Rev. Lett.* **73**, pp. 1781-1784.

Hammel, S. M., Jones, C. K. R. T. and Moloney, J. V. (1985). Global dynamical behavior of the optical field in a ring cavity, *J. Opt. Soc. Am. B* **2**, pp. 552–564.

Heagy, J. F., Carroll, T. L. and Pecora, L. M. (1994). Synchronous chaos in coupled oscillator systems. *Phys. Rev. E* **50**, pp. 1874-1885.

Hénon, M. (1976). A two-dimensional mapping with a strange attractor. *Commun. math. Phys.* **50**, pp. 69-77.

Hénon, M. (1982). On the numerical computation of Poincaré maps. *Physica D* **5**, pp. 412-414.

Haken, H. (1983). At least one Lyapunov exponent vanishes if the trajectory of an attractor does not contain a fixed point, *Phys. Lett.* A **94**, pp. 71–72.

He, R. and Vaidya, P. G. (1992). Analysis and synthesis of synchronous periodic and chaotic systems, *Phys. Rev.* A **46**, pp. 7387–7392.

Heisler, I. A., Braun, T., Zhang, Y., Hu, G. and Cerdeira, H. A. (2003). Experimental investigation of partial syncronization in coupled chaotic oscillators. *Chaos* **13**, pp. 185-194.

Herrero, R., Pons, R., Farjas, J., Pi, F. and Orriols, G. (1996). Homoclinic dynamics in experimental Shil'nikov attractors, *Phys. Rev. E* **53**, pp. 5627-5636.

Herrero, R., Figueras, M., Pi, F. and Orriols, G. (2002). Phase synchronization in bidirectionally coupled optothermal devices, *Phys. Rev. E* **66**, pp. 036223-(1-10).

Hikihara, T. and Kawagoshi, T. (1996). An experimental study on stabilization of unstable periodic motion in magneto-elastic chaos, *Phys. Lett. A* **211**, pp. 29–36.

Hindmarsh, J. L. and Rose, R. M. (1984). A model of neuronal bursting using three coupled first order differential equations. *Proc. R. Soc. Lond. B* **221**, pp 87-102.

Hu, G., Zhang, Y., Cerdeira, H. A. and Chen, S. (2000). From low-dimensional synchronous chaos to high-dimensional desynchronous spatiotemporal chaos in coupled systems. *Phys. Rev. Lett.* 85, pp. 3377-3380.

Hunt, E. R. (1991). Stabilizing high-period orbits in a chaotic system: the diode resonator, *Phys. Rev. Lett.* **67**, pp. 1953-1955.

Ikeda, K. (1979). Multiple-valued stationary state and its instability of the transmitted light by a ring cavity system, *Optics. Comm.* **30**, pp. 257-261.

Ikeda, K., Daido, H. and Akimoto, O. (1980). Optical turbulence: chaotic behavior of transmitted light from a ring cavity, *Phys. Rev. Lett.* **45**, pp. 709-712.

Jackson, L. B., *Digital Filters and Signal Processing* (Kluwer Academic Publishers, Boston, MA, 1996), pp. 462-464.

Jordan, D. W. and Smith, P., *Nonlinear Ordinary Differential Equations* (Oxford University Press, Oxford, 1990).

Kaneko, K. (1990). Clustering, coding, switching, hierarchical ordering, and control in a network of chaotic elements, *Physica D* **41**, pp. 137–172.

Kapitaniak, T. (1994). Synchronization of chaos using continuous control, *Phys. Rev.* E **50,** pp. 1642-1644.

Kapitaniak, T. (1997). Introduction, *Chaos, Solitons and Fractals* **9**, pp. xi-xii.

Kapitaniak, T. (2003). Introduction, *Chaos, Solitons and Fractals* **15**, pp. 201-203.

Kautz, R. L. (1996). Noise, chaos, and the Josephson voltage standard, *Rep. Prog. Phys.* **59**, pp. 935–992.

Kennedy, M. P. (1992). Robust op amp realization of Chua's circuit, *Frequenz* **46**, pp. 66–80.

Kennedy, M. P. and Ogorzalek, M. J. (1997). Introduction to the special isssue, *IEEE Trans. Circuits and Systems I*, **CAS-44**, pp. 853–855.

Kibble, T. W. B. and Berkshire, F. H., *Classical Mechanics* (Addison Wesley Longman, Harlow, England, 1996).

King, G. P. and Gaito S. T. (1992). Biestable chaos. I. Unfolding the cusp, *Phys. Rev.* A **46,** pp. 3092-3099.

Kiss, I. Z. and Hudson, J. L. (2001). Phase synchronization and suppression of chaos through intermittency in forcing of an electrochemical oscillator, *Phys. Rev. E* **64**, pp. 046215-(1-8).

Kiss, I. Z. and Hudson, J. L. (2002). Phase synchronization of nonidentical chaotic electrochemical oscillators, *Phys. Chem. Chem. Phys.* **4**, pp. 2638-2647.

Kittel, A., Parisi, J. and Pyragas, K. (1998). Generalized synchronization of chaos in electronic circuit experiments, *Physica D* **112**, pp. 459-471.

Kocarev, L. and Parlitz, U. (1996). Generalized synchronization, predictability, and equivalence of unidirectional coupled dynamical systems, *Phys. Rev. Lett.* **76**, pp. 1816-1819.

Krawiecki, A. and Sukiennicki, A. (2000). Generalizations of the concept of marginal synchronization of chaos, *Chaos, Solitons and Fractals* **11**, pp. 1445-1458.

Kurths, J. (2000). Guest Editorial, *Int. J. Bifurcation and Chaos* **10**, pp. 2289-2290.

Kurths, J., Boccaletti, S., Grebogi, C. and Lai, Y.-C. (2003). Introduction: control and synchronization in chaotic dynamical systems, *Chaos* **13**, pp. 126-127.

Lima, R. and Pettini, M. (1990). Suppression of chaos by resonant parametric perturbations, *Phys. Rev.* A **41**, pp. 726–733.

Liu, Y., de Oliveira, P. C., Danailov, M. B. and Rios Leite, J. R. (1994). Chaotic and periodic passive Q switching in coupled CO_2 lasers with saturable absorber, *Phys. Rev. A* **50**, pp. 3464-3470.

Lee, K. J., Kwak, Y. and Lim, T. K. (1998). Phase jumps near a phase synchronization transition in systems of two coupled chaotic oscillators, *Phys. Rev. Lett.* **81**, pp. 321-324.

Lorenz, E. N. (1963). Deterministic nonperiodic flow, *J. Atmos. Sci.* **20**, pp. 130–141.

Lorenz, E. N. (1984). Irregularity: a fundamental property of the atmosphere, *Tellus A* **36**, pp. 98–110.

Lüthje, O., Wolff, S. and Pfister, G. (2001). Control of chaotic Taylor-Couette flow with time-delayed feedback, *Phys. Rev. Lett.* **86**, pp. 1745–1748.

Mackey, M. C. and Glass, L. (1977). Oscillations and chaos in physiological control systms, *Science* **197**, pp. 287-289.

Mainieri, R. and Rehacek, J. (1999). Projective synchronization in three-dimensional chaotic systems. *Phys. Rev. Lett.* **82**, pp. 3042-3045.

Mandelbrot, B. B., *The fractal geometry of nature* (W. H. Freeman and Company, New York, 1983).

Masoller, C. (2001). Anticipation in the synchronization of chaotic semiconductor lasers with optical feedback, *Phys. Rev. Lett.* **86**, pp. 2782-2785.

Matías, M. A., Güémez, J. and Martín, C. (1997). On the behavior of coupled chaotic systems esxhibiting marginal synchronization. *Phys. Lett.* A **226,** pp. 264-268.

Matsumoto, T., Chua, L. O. and Komuro, M. (1985). The Double Scroll, *IEEE Trans. Circuits and Systems,* **CAS-32**, pp. 798–818.

May, R. M. (1976). Simple mathematical models with very complicated dynamics. *Nature* **261**, pp. 459–467.

McCoy, J. K., Parmananda, P., Rollins, R. W. and Markworth, A. J. (1993). Chaotic dynamics in a model of metal passivation, *J. Mater. Res.* **8**, pp. 1858-1865.

Mesulam, M. M. (1990). Large-scale neurocognitive networks and distributed processing for attention, language, and memory. *Ann. Neurol.* **28**, pp. 597–613.

Miranda, R. and Stone, E. (1993). The proto-Lorenz system, *Phys. Lett. A* **178**, pp.105-113.

Moon, F. C., *Chaotic vibrations* (John Wiley & Sons, New York, 1987).

Newcomb, R. W. and Sathyan, S. (1983). An RC op amp chaos generator, *IEEE Trans. Circuits and Systems*, CAS-30, pp. 54–56.

Newhouse, S., Ruelle, D. and Takens, F. (1978). Occurrence of strange axiom A attractors near quasi periodic flows on $T^m, m \geqq 3$ *Commun. math. Phys.* **64**, pp. 35–40.

Nicolis, G. (1990). Chemical chaos and self-organization. *J. Phys.: Condens. Matter* **2**, pp. SA47–SA62.

Osipov, G. V., Pikovsky, A. S., Rosenblum, M. G. and Kurths, J. (1997). Phase synchronization effects in a lattice of nonidentical Rössler oscillators, *Phys. Rev. E* **55**, pp. 2353-2361.

Osipov, G. V., Hu, B., Zhou C., Ivanchenko, M. V. and Kurths, J. (2003). Three types of transitions to phase synchronization in coupled chaotic oscillators, *Phys. Rev. Lett.* **91**, pp. 024101-(1-4).

Ott, E., Grebogi, C. and Yorke, J. A. (1990). Contolling chaos, *Phys. Rev. Lett.* **64**, pp. 1196-1199.

Packard, N. H., Crutchfield, J. P., Farmer, J. D. and Shaw, R. S. (1980). Geometry from a time series, *Phys. Rev. Lett.* **45**, pp. 712–716.

Park, E.-H., Zaks, M. A. and Kurths, J. (1999). Phase synchronization in the forced Lorenz system. *Phys. Rev. E* **60**, pp. 6627-6638.

Parmananda, P., Madrigal, R., Rivera, M., Nyikos, L., Kiss, I. Z. and Gáspár, V. (1999). Stabilization of unstable steady states and periodic orbits in an electrochemical system using delayed-feedback control. *Phys. Rev. E* **59**, pp. 5266-5271.

Pawelzik, K. and Schuster, H. G. (1991). Unstable periodic orbits and prediction. *Phys. Rev. A* **43**, pp. 1808-1812.

Pecora, L. M. and Carroll, T. L. (1990). Synchronization in chaotic systems. *Phys. Rev. Lett.* **64**, pp. 821-824.

Pecora, L. M. and Carroll, T. L. (1991). Driving systems with chaotic signals. *Phys. Rev. A.* **44**, pp. 2374-2383.

Pecora, L. M., Carroll, T. L., Johnson, G. A., Mar, D. J. and Heagy, J. F. (1997). Fundamentals of synchronization in chaotic systems, concepts, and applications. *Chaos* **7**, pp. 520-543.

Pecora, L. M. and Carroll, T. L. (1998). Master Stability Functions for Synchronized Coupled Systems. *Phys. Rev. Lett.* **80**, pp. 2109-2112.

Peng, B., Petrov, V. and Showalter, K. (1991). Controlling chemical chaos, *J. Phys. Chem.* **95**, pp. 4957-4959.

Pérez-Muñuzuri, V., Matías, M. A. and Kapral, R. (1999). Foreword, *Int. J. Bifurcation and Chaos* **9**, pp. 2127-2128.

Petrov, V., Gáspár, V., Masere, J. and Showalter, K. (1993). Controlling chaos in the Belousov-Zhabotinsky reaction, *Nature.* **361**, pp. 240-243.

Pikovsky, A. S., Rosenblum, M. G. and Kurths, J. (1996). Synchronization in a population of globally coupled chaotic oscillators, *Europhys. Lett.* **34**, pp. 165–170.

Pikovsky, A. S., Rosenblum, M. G., Osipov, G. V. and Kurths, J. (1997a). Phase synchronization of chaotic oscillators by external driving, *Physica* D **104**, pp. 219–238.

Pikovsky, A., Zaks, M., Rosenblum, G., Osipov, G. and Kurths, J. (1997b). Phase synchronization of chaotic oscillators in terms of periodic orbits, *Chaos* **7**, pp. 680–687.

Pikovsky, A., Rosenblum, G. and Kurths, J. (2000). Phase synchronization in regular and chaotic systrems, *Int. J. Bifurcation and Chaos* **10**, pp. 2291–2305.

Pingel, D., Schmelcher, P. and Diakonos, F. K. (2001). Detecting Unstable Periodic orbits in chaotic continuous-time dynamical systems. *Phys. Rev. E* **64**, pp. 026214-(1-10).

Press, W. H., Teukolsky, S. A., Vetterling, W. T. and Flannery, B. P., *Numerical Recipes in Fortran 77: The Art of Scientific Computing* (Cambridge University Press, Cambridge, UK, 1992).

Pujol-Peré, A., Calvo, O., Matías, M. A. and Kurths, J. (2003). Experimental study of imperfect phase synchronization in the forced Lorenz system *Chaos* **13**, pp. 319-326.

Pyragas, K. (1992). Continuous control of chaos by self-contolling feedback. *Phys. Lett.* **170**, pp. 421-428.

Pyragas, K. (1996). Weak and strong synchronization of chaos. *Phys. Rev. E* **54**, pp. R4508-R4511.

Pyragas, K. (2002). Analytical properties and optimization of time-delayed feedback control. *Phys. Rev. E.* **66**, pp. 026207-(1-9).

Qu, Z., Hu, G., Yang, G. and Qin, G. (1995). Phase effect in taming nonautonomous chaos by weak harmonic perturbations. *Phys. Rev. Lett.* **74**, pp. 1736-1739.

Rabinovich, M. I., Abarbanel, H. D. I., Huerta, R., Elson, R. and Selverston, A. I. (1997). Self-regularization of chaos in neural systems: experimental and theoretical results, *IEEE Trans. CAS.* **44**, pp. 997-1005.

Rajasekar, S., and Lakshmanan, M. (1994). Bifurcation, chaos and suppression of chaos in Fitz-Hugh-Nagumo Nerve Conduction Model Equation, *J. theor. Biol.* **166**, pp. 275-288.

Riley, K. F., Hobson, M. P. and Bence, S.J., *Mathematical methods for Physics and Engineering* (Cambridge University Press, Cambridge, UK, 1998).

Roebber, P. J. (1995). Climate variability in a low-order coupled atmosphere-ocean model, *Tellus A* **47**, pp. 473–494.

Rosa Jr., E., Ott, E. and Hess, M. H. (1998). Transition to Phase Synchronization of Chaos, *Phys. Rev. Lett.* **80**, pp. 1462–1465.

Rosa Jr., E., Pardo, W. B., Ticos, C. M., Walkenstein, J. A. and Monti, M. (2000). Phase synchronization of chaos in a plasma discharge tube *Int. J. Bifurcation and Chaos* **10**, pp. 2551-2563.

Rosenblum, M. G., Pikovsky, A. S. and Kurths, J. (1996). Phase synchronization of chaotic oscillators, *Phys. Rev. Lett.* **76**, pp. 1804–1807.

Rosenblum, M. G., Pikovsky, A. S. and Kurths, J. (1997). From phase to lag synchronization in coupled chaotic oscillators, *Phys. Rev. Lett.* **78**, pp. 4193–4196.

Rössler, O. E. (1976). An equation for continuous chaos, *Phys. Lett. A* **57**, pp. 397–398.

Rössler, O. E. (1979). An equation for hyperchaos, *Phys. Lett. A* **71**, pp. 155–157.

Roy, R., Murphy Jr., T. W., Maier, T. D., Gills, Z. and Hunt, E. R. (1992). Dynamical control of a chaotic laser: experimental stabilization of a globally coupled system, *Phys. Rev. Lett.* **68**, pp. 1259-1262.

Roy, R. and Thornburg Jr., K. S. (1994). Experimental synchronization of chaotic lasers, *Phys. Rev. Lett.* **72**, pp. 2009–2012.

Ruelle D. and Takens, F. (1971). On the nature of turbulence, *Commun. math. Phys.* **20**, pp. 167–192.

Rul'kov, N. F., Volkovskii, A. R., Rodríguez-Lozano, A., del Rio, E. and Velarde, M. G. (1992). Mutual synchronization of chaotic self-oscillators with dissipative coupling, *Int. J. Bifurcation and Chaos* **2**, pp. 669-676.

Rulkov, N. F., Sushchik, M. M., Tsimring, L. S. and Abarbanel, H. D. I. (1995). Generalized synchronization of chaos in directionally coupled chaotic systems. *Phys. Rev. E* **51**, pp. 980-994.

Rulkov, N. F. and Afraimovich, V. S. (2003). Detectability of nondifferentiable generalized synchrony. *Phys. Rev. E* **67**, pp. 066218-(1-8).

Rybski, D., Havlin, S. and Bunde, A. (2003). Phase synchronization in temperature and precipitation records, *Physica A* **320**, pp. 601–610.

Sauer, T., Yorke, J. A. and Casdagli, M. (1991). Embedology, *J. Stat. Phys.* **65**, pp. 579–616.

Savage, H. T., Ditto, W. L., Braza, P. A., Spano, M. L., Rauseo, S. N. and Spring, W. C. (1990). Crisis-induced intermittency in a parametically driven, gravitationally buckled, magnetoelastic amorphous ribbon experiment, *J. Appl. Phys.* **67**, pp. 5619–5623.

Schiff, S. J., Jerger, K., Duong, D. H., Chang, T., Spano, M. L. and Ditto, W. L. (1994). Contolling chaos in the brain, *Nature* **370**, pp. 615-620.

Schiff, S. J., So, P., Chang, T., Burke, R. E. and Sauer, T. (1996). Detecting dynamical interdependence and generalized synchrony through mutual prediction in a neural ensemble. *Phys. Rev. E* **54**, pp. 6708–6724.

Schmelcher, P. and Diakonos, F. K. (1997). Detecting unstable periodic orbits of chaotic dynamical systems, *Phys. Rev. Lett.* **78**, pp. 4733-4736.

Schreiber, T. (1999). Interdisciplinary application of nonlinear time series methods, *Phys. Rep.* **308**, pp. 1-64.

Scott, A. C. (1975). The electophysics of a nerve fiber, *Rev. Mod. Phys.* **47**, pp. 487-533.

Shil'nikov, A., Nicolis, G. and Nicolis, C. (1995). Bifurcation and predictability analysis of a low-order atmospheric circulation model, *Int. J. Bifurcation and Chaos* **5**, pp. 1701–1711.

Shinbrot, T., Ott, E., Grebogi, C. and Yorke, J. A. (1990). Using chaos to direct trajectories to targets, *Phys. Rev. Lett.* **65**, pp. 3215-3218.

Shinbrot, T., Grebogi, C., Ott, E. and Yorke, J. A. (1993). Using small perturbations to control chaos, *Nature* **363**, pp. 411-417.

Shimada, I., and Nagashima, T. (1979). A numerical approach to ergodic problem of dissipative dynamical systems, *Prog. Theor. Phys.* **61**, pp. 1605–1616.

Shinriki, M., Yamamoto, M. and Mori, T. (1981). Multimode oscillations in a modified van der Pol oscillator containing a positive nonlinear conductance, *Proc. IEEE* **69**, pp. 394–395.

Singer, J., Wang, Y.-Z. and Bau, H. H. (1991). Controlling a chaotic system, *Phys. Rev. Lett.* **66**, pp. 1123-1125.

Sivaprakasam, S. and Shore, K. A. (1999). Demonstration of optical synchronization of chaotic external-cavity laser diodes, *Opt. Lett.* **24**, pp. 466–468.

Socolar, J. E. S., Sukow, D. W. and Gauthier, D. J. (1994). Stabilizing unstable periodic orbits in fast dynamical systems, *Phys. Rev. E.* **50**, pp. 3245–3248.

Sprott, J. C. (1994) Some simple chaotic flows, *Phys. Rev. E.* **50**, pp. R647–R650.

Stefanski, A., Kapitaniak, T. and Brindley, J. (1996). Dynamics of coupled Lorenz systems and its geophysical implications, *Physica D* **98**, pp. 594-598.

Sugawara, T., Tachikawa, M., Tsukamoto, T. and Shimizu, T. (1994). Observation of synchronization in laser chaos, *Phys. Rev. Lett.* **72**, pp. 3502–3505.

Swinney, H. L. (1983). Observations of order and chaos in nonlinear systems, *Physica D* **7**, pp. 3–15.

Szebehely, V. (1984). Review of conceps of stability, *Celestial mech.* **34**, pp. 49-64.

Thaherion, S. and Lai, Y.-C. (1999). Observability of lag synchronization of coupled chaotic oscillators, *Phys. Rev. E.* **59**, pp. R6247-R6250.

Takens, F., Detecting strange attractors in turbulence, in *Lecture Notes in Mathematics* **898** (Springer, Berlín 1981).

Tang, D. Y., Dykstra, R., Hamilton, M. W. and Heckenberg, N. R. (1998). Observation of generalized synchronization of chaos in a driven chaotic system, *Phys. Rev. E* **57**, pp. 5247–5251.

Tang, S., and Liu, J. M. (2003). Experimental verification of anticipated and retarded synchronization in chaotic semiconductor lasers, *Phys. Rev. Lett.* **90**, pp. 194101-(1-4).

Tass, P., Rosenblum, M. G., Weule, J., Kurths, J., Pikovsky, A., Volkmann, J., Schnitzler, A. and Freund, H.-J. (1998). Detection of n:m phase locking from noisy data: application to magnetoencephalography, *Phys. Rev. Lett.* **81**, pp. 3291–3294.

Terry, J. R. and Breakspear, M. (2003). An improved algorithm for the detection of dynamical interdependence in bivariate time series, *Biol. Cybern.* **88**, pp. 129–136.

Ticos, C. M., Rosa Jr., E., Pardo, W. B., Walkenstein, J. A. and Monti, M. (2000). Experimental Real-Time Phase Synchronization of a Paced Chaotic Plasma Discharge, *Phys. Rev. Lett.* **85**, pp. 2929–2932.

Tsonis, A. A. (2001). The impact of nonlinear dynamics in the atmospheric sciences, *Int. J. Bifurcation and Chaos* **11**, pp. 881-902.

Uchida, A., Shinozuka, M., Ogawa, T. and Kannari, F. (1999). Experiments on chaos synchronization in two separate microchip lasers, *Opt. Lett.* **24**, pp. 890–892.

VanWiggeren, G. D. and Roy, R. (1998). Communication with chaotic lasers, *Nature* **279**, pp. 1198-1200.

Varela, F., Lachaux, J.-P., Rodriguez, E. and Martinerie, J. (2001). The brainweb: phase synchronization and large-scale integration. *Nature Reviews Neuroscience* **2**, pp. 229–239.

Voss, H. U. (2000). Anticipating chaotic synchronization, *Phys. Rev. E.* **61**, pp. 5115-5119.

Voss, H. U. (2001). Dynamic Long-Term Anticipation of Chaotic States, *Phys. Rev. Lett.* **87**, pp. 014102-(1-4).

Voss, H. U. (2002). Real-time anticipation of chaotic states of an electronic circuit, *Int. J. Bifurcation and Chaos* **12**, pp. 1619-1625.

Wang, W., Kiss, I. Z. and Hudson, J. L. (2000). Experiments on arrays of globally coupled chaotic electrochemical oscillators: Synchronization and clustering. Chaos 10, pp. 248-256.

Wedekind, I. and Parlitz, U. (2002) Synchronization and antisynchronization of chaotic power drop-outs and jump-ups of coupled semiconductor lasers, *Phys. Rev. E.* **66**, pp. 026218–(1-4).

Winfree, A. T., *The Geometry of Bilogical Time* (Springer-Verlag, New York, 1980).

Wolf, A., Swift, J. B., Swinney, H. L. and Vastano, J. A. (1985). Determining Lyapunov exponents from a time series, *Physica D* **16**, pp. 285–317.

Xu, D., Li, Z.. and Bishop, S. R. (2001). Manipulating the scaling factor of projective synchronization in three-dimensional chaotic systems, *Chaos* **11**, pp. 439–442.

Xu, D., Ong, W. L. and Li, Z. (2002). Criteria for the occurrence of projective synchronization in chaotic systems of arbitrary dimension, *Phys. Lett. A* **305**, pp. 167–172.

Yalçinkaya, T. and Lai, Y.-C. (1997). Phase characterization of chaos, *Phys. Rev. Lett.* **79**, pp. 3885–3888.

Zaks, M. A., Park, E.-H., Rosenblum, M. G. and Kurths, J. (1999). Alternating Locking Ratios in Imperfect Phase Synchronization, *Phys. Rev. Lett.* **82**, pp. 4228–4231.

Zhabotinsky, A. M. (1991). A history of chemical oscillations and waves, *Chaos* **1**, pp. 379–386.

Zhu L. and Lai, Y.-C. (2001). Experimental observation of generalized time-lagged chaotic synchronization, *Phys. Rev. E* **64**, pp. 045205–(1-4).

Index